Elements of
ANIMAL PHYSIOLOGY

Elements of ANIMAL PHYSIOLOGY

M. Nithya Devi

ANMOL PUBLICATIONS PVT. LTD.
NEW DELHI-110 002 (INDIA)

ANMOL PUBLICATIONS PVT. LTD.
Regd. Office: 4360/4, Ansari Road, Daryaganj,
New Delhi-110002 (India)
Tel.: 23278000, 23261597, 23286875, 23255577
Fax: 91-11-23280289
Email: anmolpub@gmail.com
Visit us at: www.anmolpublications.com

Branch Office: No. 1015, Ist Main Road, BSK IIIrd Stage
IIIrd Phase, IIIrd Block, Bengaluru-560 085 (India)
Tel.: 080-41723429 • Fax: 080-26723604
Email: anmolpublicationsbangalore@gmail.com

Elements of Animal Physiology

First Edition, 2011

PRINTED IN INDIA

Printed at Mehra Offset Press, Delhi.

Contents

Preface

Animal Physiology is the essential core text for all those studying physiology or zoology. The advances that have taken place in the field of physiology during the last four to five decades are spectacular. It is therefore necessary that the student either at undergraduate or postgraduate level is fully exposed to these frontiers in Animal Physiology and the present publication is an attempt in this direction. To keep pace with the rapid advances in the health sciences, today's health professional students must be able to draw on their conceptual understanding of physiology instead of merely recalling isolated facts that soon may be outdated. This book provides extensive account of recent developments in the field. Attempts have been made to design a textbook by incorporating important topics that are useful and effective for students.

Author

1

Animal Morphology

Animal Anatomy, is the study of the shape, form, and structure of animals and their parts. Morphologists are interested in questions such as how the fang mechanisms of cobras differ from those of rattlesnakes or how the wings of bats evolved from the forelimbs of their ancestors. Animals can be broadly distinguished into two major groups the nonchordates (or invertebrates) and the chordates. Invertebrates account for more than 95 per cent of the animal species that have been identified. These are animals which do not have a vertebral column or a back bone.

They are distributed into nine major phyla from protozoans to echinoderms. The chordates are animals which have a notochord on the dorsal side of the body. Most of the chordates can be described as vertebrates since the notochord in these animals is restricted to the embryonic stage and in the adults it becomes replaced by a vertebral column. A very distinct evolutionary trend in the general organisation of animals can be noticed as one proceeds from protozoans to chordates. From an acellular level of organisation to cellular level and then tissue level, organ level and organ system level of organization can be recognised. In addition, one can recognise the gradual appearance of distinct features such as symmetry, body wall and body cavity.

There are further improvements in the form of appearance of segmentation, body divisions and even formation of a distinct head (cephalistion). The culmination of all these evolutionary advances, is seen in the mammals which represent the most recent group of animals to have evolved and at the same time the most advanced group of animals. In order to gain an insight into the evolutionary advances among animals, it become necessary to study the salient features of representative animals from the group of invertebrates and vertebrates. If one examines a living cell under the highest powers of the microscope one sees surprisingly little. In view of the fact that protoplasm is capable of complicated physical and chemical transformations it might be supposed that it would posses intricate mechanical arrangements.

As everyone knows, the visible structure of a living cell is rather simple, and there is nothing to suggest the complexities of man-made machines. This may mean that the fundamental structure of protoplasm is ultramicroscopic or in the range of molecular dimensions. Following the establishment of the cell doctrine the science of cytology was built up mostly along morphological lines.

Cytologists sought to find in the structure of the cell some explanation of its life. In their study of the nucleus they were to some extent successful. Some years ago, the study of the huge chromosomes found in the cells of the salivary glands of fly larvae completely confirmed the earlier work of T. H. Morgan and his school, and showed that the genes are arranged in linear fashion along the chromosomes. Thus the visible structure of the chromosome bears a relation to its function in heredity. The position of the genes is important in determining the way in which they are distributed under various conditions, but the fact that they are arranged in linear order tells us nothing as to how a gene can influence the development of one character or another. In their study of the structure of cytoplasm cytologists have not been so fortunate.

Although the cell nucleus is of primary importance in inheritance, for the ordinary routine life of the cell the cytoplasm plays the leading role. This is indicated by various experiments in which cells were deprived of nuclei. Thus, Lynch, among others, divided amebae into halves, one with nucleus, the other without. The fragment without a nucleus was able to live for days, although apparently it was incapable of synthesizing nitrogenous compounds. Similar studies of other workers indicate that cells without nuclei can live for a time. A surprising case is that of Acetabularia mediterranea, a large unicellular alga which posses but a single nucleus that is easy to remove.

In this form, cytoplasm without nucleus can live for months and is capable of regenerating some rather complicated structures. On the other hand, in the nerve cells of higher animals, growth of cytoplasm can occur only in the region of the nucleus. Studies of the structure of cytoplasm have yielded very little which has proved of much help in the interpretation of its function. Perhaps because of this, some modern cytologists seem to be losing interest in the classical cytology which dealt almost entirely with structure and have become more and more concerned with the chemistry and the physiology of the cell.

All living cells are believed to be surrounded by definite membranes. The evidence for this statement has come almost entirely from the physiological side. Since the early days of cytology morphologists have been but little concerned with the outer boundaries of cells. If one examines a sea urchin egg carefully one can see that it is bounded by a delicate membrane perhaps 1μ in thickness. This membrane is very hard to distinguish, for it appears to have the same refractive index as the sea water surrounding the egg. Whether the membrane which is visible around the sea urchin egg is the same membrane as that which governs the osmotic behaviour of the cell is not certain.

The simplest theory would be to assume that it is. However, some authors insist that osmotic membranes surrounding cells are much thinner than the membrane which can be seen to envelop the sea urchin egg. It is possible that the cell boundaries or cell walls which can be so clearly seen in preparations of animal tissues are really membranes of physiological importance. The question as to whether they are or not has scarcely been discussed. Most biologists are influenced by the relation one finds in plant tissues. The cell walls of plant cells are non-living, and serve merely as supports for the protoplasmic structures within. On being placed in concentrated solutions plant cells shrink away from their walls. There is evidence that, within the outer membrane of the animal cell, there is a cortex, different both in morphological and physical properties from the cell interior.

In some protozoan cells such a cortex can be clearly recognized. In metazoan cells it is not so easy to distinguish. However, Just has presented both morphological and physiological evidence to show the presence of a cortex as a very important part of the metazoan cell. Within the outer membrane and the cortex of a living cell lies the main bulk of the cytoplasm. Cytologists have struggled hard to discover structures in the cytoplasm which would help to explain its activity. Because of the fact that living cells reveal relatively little, various methods of fixing, sectioning, and staining have been employed.

Cells, or groups of cells, are treated with strong acids, salts of heavy metals, alcohol or formaldehyde, or with various combinations of these and other reagents. Following fixation the cells are bathed in alcohol, then in some solvent of paraffin, and then (typically) they are impregnated with paraffin, cut into sections, and stained. These methods of treatment have been very successful in the study of chromosomes, and it can be shown by comparison with the living that the chromosomes are not markedly altered by all these procedures. The value of fixation and staining in the study of cytoplasm is less certain. Fischer showed at great length that fixation agents caused the production of various artifacts in protein solutions, and that these artifacts resembled the granules and filaments described by cytologists as occurring in living cells. Indeed, if one watches the action of a fixing agent on living cells, one may, under proper conditions, observe the formation of new granules or vacuoles, or even the disappearance of granules normally present.

Since the publication of Fischer's book cytologists have been more careful to compare the structures they see in fixed and stained material with the living protoplasm. In the latter half of the nineteenth century students of the cell urged one type of structure or another as essential for the living protoplasm. Thus, Frommann, Heitzmann, Klein and Leydig, all argued that protoplasm consisted of a net of solid material, the meshes of which were filled with fluid. Flemming believed that fibrils were the most important constituent of the protoplasm. On the other hand, Altmann vigourously maintained the importance of granules. He believed that the granules represented minute organisms living within a more or less inanimate fluid.

Finally, the protozoologist, Bütschli, thought of protoplasm as a foam, or rather as an emulsion composed of 2 liquids, one in the form of droplets, the other as lamellae between the droplets. These nineteenth century views of the fundamental nature of living substance are not as new as is generally supposed. Even in the eighteenth century many observers had noted the presence of innumerable granules in living material, and there was also some talk of the existence of fibrils. The twentieth century saw a new development, or rather a variation of the old one. In 1897, Benda coined the word *mitochondria* for granules which he found first in rat spermatocytes and later in other types of cells. These mitochondria often showed a tendency to unite in the form of threads. Benda believed that his complicated technic was specific for the mitochondria, and that these represented a new discovery.

Actually, mitochondria are not to be distinguished from the granules of Altmann or the filaments of Flemming. It is important to note that not all the granules of the cytoplasm are to be considered as mitochondria. Nor is it always entirely clear just which granules are mitochondria and which ones are not. In general mitochondria are coloured with the vital dye Janus green, and Sorokin regards Janus green B as "virtually specific" for mitochondria. They can be demonstrated by the use of those fixatives which do not dissolve lipids, and it is believed that chemically mitochondria consist of a mixture of protein with a phospholipid (*e.g.*, lecithin). Mitochondria can be seen in living protoplasm, and they are believed to be necessary constituents of all plant and animal cells. In plant cells mitochondria may become transformed into plastids. In animal cells they have been claimed to transform into secretion granules, fibrils, and various other structures of differentiated cells. Concerning their physiological function little has been known until recently. Now it is believed that mitochondria contain oxidizing enzymes, especially the enzymes known as succinoxidase and cytochrome oxidase. Soon after the discovery of mitochondria, or rather soon after the introduction of the concept by Benda, another type of cytoplasmic structure was introduced to the cytological world.

Studying the nerve cells of an owl, Golgi described the presence of a network, the so-called "Golgi apparatus." This may be in the form of a net, or it may be in the form of granules or rods. It is demonstrated either by a silver impregnation process or by long continued soaking in osmic acid (OsO_4) solutions. For many years the Golgi apparatus was believed to have a great significance, for it could be demonstrated in many diverse types of protoplasm; and also, according to a number of authors, it could be seen in various types of living cells. Some workers have found that the region of the cell which contains the Golgi apparatus is rich in phosphatase enzymes, that is, enzymes which are able to split phosphate from organic combination.

However, in recent years there has been growing scepticism concerning the x. According to Hibbard, a leading worker in the field, "there does not seem to be much agreement as to the nature and functioning of the Golgi apparatus" and she feels that in so far as most of the generalizations that have been made are concerned, "we should be better off if the discovery of the impregnation methods had never been made." From time to time, it has been urged that the Golgi apparatus as such does not exist in living cells at all, but is an artifact produced by the reagents which are used to demonstrate it. Palade and Claude have presented forcible arguments in support of this view, but these arguments have been vigourously disputed. Finally, many cells contain vacuoles, which may be minute or large. A comprehensive term for the vacuoles of a cell is "vacuome." Vacuoles may appear and disappear. They can originate from mitochondria, or be quite distinct from them. Sometimes they are produced by reagents, as will be noted later. Occasionally cells are so filled with vacuoles that the protoplasm has the appearance of a foam. This type of protoplasm is most common in the protozoa, and it is easy therefore to understand why Bütschli regarded all protoplasm as foamlike. Students of the finer structures of cells have always attempted to use microscopes which show the greatest possible detail. For this purpose, the limiting factor is not the extent to which the microscopic image may be enlarged, for the degree of possible enlargement is infinite.

The limiting factor is the "resolving power," or the power of discriminating between 2 minute objects that lie close together. Years ago, the resolving power of ordinary microscopes was developed to its very greatest extent, for the resolving power is primarily dependent on the wave length of the light passing through the lenses of the microscope and is actually inversely proportional to this wave length. By the use of ultraviolet radiation, which has shorter wave lengths than visible light, a 2 1/2 fold increase in resolving power could be obtained, and ultraviolet microscopes were developed. In these, quartz lenses were employed, inasmuch as glass is opaque to ultraviolet radiation. A far greater increase in resolving power may be obtained if instead of visible light or ultraviolet radiation, cathode rays (beams of electrons) are employed. Modern physics regards the electron as a particle having a wave nature, and the wave length of such radiation may be only 1/100,000th that of visible light.

Hence it might be possible to obtain vastly greater resolving power than that possible with the ordinary microscope. Developments in the new science of electron optics have made possible the construction of electron microscopes. In such microscopes the electrons are focused by electron lenses, which consist of electrostatic or magnetic fields capable of deflecting the electrons. Cathode rays are effective only in a vacuum, and biological material to be studied in an electron microscope must be thoroughly dried. The drying process itself can cause violent distortion. Thus if some cells are suspended in a droplet of water and this is then allowed to evaporate, the surface of the water eventually encounters the cells. Surface tension exerts relatively enormous forces on small objects and the cells may be badly crushed. To avoid such surface effects, Anderson has found it desirable to replace the water with liquid carbonic acid which is gaseous at ordinary temperatures, and then to evaporate at temperatures above the critical temperature of the gas. In this way, better pictures can be obtained.

Of course the electrons or cathode rays themselves are destructive so that living material even if it were to survive the drying would be killed by the cathode rays. In spite of all these shortcomings, the electron microscope has been able to demonstrate within bacteria structures not visible with the ordinary microscope, and it is extremely useful in the study of bacteriophages and viruses. In order to examine larger cells under the electron microscope they must either be pulled apart, which can cause violent change, or they must be studied in thin sections.

For the preparation of very thin sections, special microtomes have been constructed. So far the study of large cells with the electron microscope has not yielded results of any great interest or importance. The morphological study of the cytoplasm has resulted in a large body of information concerning the various granular and other formed constituents of living substance. Unfortunately, however, there are no very definite clues as to the relation of these formed elements to the vital activity. One point seems certain: the granules or filaments of the cytoplasm are not present in any spatial arrangement of significance for the physiology of the cell.

At least this is true for the main mass of the protoplasm, although it may not be true for the cortex. When cells are centrifuged, even at very high speeds, the distribution of the granules may be profoundly altered without affecting the life or subsequent behaviour of the cell. It should be noted that the granules of living cells may be present in tremendous numbers. In the egg of the sea urchin Arbacia, the granules occupy nearly 20 per cent of the total volume. In the egg of the starfish Asterias there is a much higher concentration of granules, and the egg of the sea urchin Paracentrotus is so full of granules that there is hardly any free space between them. Paramecium protoplasm is also densely packed with granules. The presence of so many granules must have an important effect on the physical properties of the protoplasm as a whole.

CHEMICAL CONSTITUENTS OF LIVING ORGANISMS

More than is often realised, scientific progress is dependent on the coining of apt phrases and catchwords. The study of granules in cells was certainly fostered by Benda's introduction of the word "mitochondria." So too the introduction of the word "vitamin" was a factor in the development of a field which had never prospered as long as vitamins were called by such awkward terms as "accessory articles of the diet."

In the study of living material the concept of a cell theory or a cell doctrine has attracted the attention of students and workers to the fact that even the most diverse types of animals and plants are constituted of the same type of unit. It is probably true that living material is as basically similar in chemical composition as in microscopical appearance; and yet this important fact, unheralded by any catchword, has never received the emphasis it deserves. True, the concept of protoplasm has had a wide influence, but the term "protoplasm" is commonly used more in a morphological sense than in a chemical one.

At the present time, indeed, one is not quite sure as to just how constant protoplasm is in a chemical sense. Numerous analyses have been made of the organs, tissues, and cells of living organisms, but these studies have been made by biochemists who have generally had little or no interest in the protoplasm concept, and were not concerned with the degree of uniformity shown by diverse types of living material. Just as the cytologist has attempted to explain life by a study of the minute details of microscopic anatomy of cells, so the chemicallyminded biologist might seek to explain life by studying the chemical composition of living substance. As a matter of fact, however, students of the chemistry of living material have never been very optimistic in this direction. Obviously, almost all types of chemical analysis first lead to death of the cell, so that with ordinary chemical methods one can only obtain information regarding that which *was* once alive. Indeed, the chemical properties of living protoplasm are but little understood. We do not clearly understand what it is that distinguishes living protoplasm from dead, although it is certain that chemical changes do accompany and follow the death process.

Chemically, it is possible to isolate from living material various types of organic compounds. Usually, it is fair to assume that the compounds isolated were present as such when the protoplasm was alive. And yet there is good reason to believe that frequently substances obtained from cells were really in life parts of more complex compounds. So, for example, a carbohydrate extracted from living tissue may before the extraction have been bound to a protein; similarly fats or fat-like substances gotten from cells may in the living cells have been tied to protein.

Also, individual proteins which are obtained in extracts of tissues may originally have been combined with other proteins. This is now well known to be true in the case of muscle. Thus it appears that all the organic compounds known to constitute the mass of the living protoplasm may to a large extent be tied together. Also they may be tied to inorganic constituents such as phosphoric acid, sulfate, metals, alkaline earth cations or even water. One of the most interesting aspects of the chemistry of living material is the ability of the chemist to extract from it certain catalytic substances or enzymes.

These enzymes can be isolated in rather pure form and their effect on chemical reactions can be followed in test tubes. But even in the case of these enzymes, it is not certain that the extracted enzymes are in the same state as they were in the cell. Thus von Euler and von Euler urge that within the cell enzymes may exist in the form of complexes. Following extraction, the complexes are broken up into simpler constituents. If so many substances of a cell are tied together, we might be inclined to think that protoplasm consisted in reality of giant molecules containing practically everything of importance to the life of the cell.

Such ideas have been held in the past. For example, Verwor n conceived of a giant protoplasmic molecule, which he called biogen. Verworn's ideas of the nature of biogen could hardly be held now, and indeed the modern biochemist prefers to think of protoplasm as composed of many organic compounds existing separately within the cell. If, however, at some future time viruses come to be regarded as truly living, we might have to extend our idea of protoplasm so as to include materials consisting of just one type of chemical compound.

ELEMENTS IN PROTOPLASM

There may be uncertainty as to exactly which compounds exist within the living protoplasm, or as to how the various compounds are tied together, but it is a relatively simple matter to determine exactly which elements are present and in what amount. The ordinary methods that the chemist uses for qualitative and quantitative analysis can be employed. In addition, spectographic methods are very helpful, and they can be used for minute amounts of material. In the newer spectrography, quantitative determination of elements has been made possible by the use of photoelectric devices. The radiation emitted by an ignited substance is split up into its constituent wave lengths and the spectrum may then be recorded on photographic film or plate.

Then the intensities of the spectral lines are determined by means of a densitometer or microphotometer and these intensities are compared with a calibration curve obtained from known substances in known amounts. Or without photographic film one can compare the intensity of spectral emission lines with the intensity of the lines given by known quantities of known elements. The flame photometer is now widely used in physiological laboratories and is especially suitable for the rapid determination of potassium and sodium. In this apparatus, a solution containing sodium, potassium or other elements is fed into a flame and the emission spectrum is then compared with that of a solution containing a known amount of the element. However, it must not be supposed that the determination of elements within living material is always easy. There are difficulties which are often not fully realised. True, standard methods of quantitative analysis are applicable, but such methods can be properly used only if the material to be examined is in true solution. Usually, moreover, inorganic elements to be recognized must be in ionic form.

It is not always an easy matter to dissolve living substance completely and to free it from carbon and from other substances which would interfere with the reactions of quantitative analysis. In order to proceed with such an analysis, the living tissue must first be ashed. Because of the fact that elements like chlorine and potassium volatilize at not too high temperatures, most biochemists use what are called wet-ashing procedures. The tissue is heated with concentrated nitric or sulfuric acid or both, and in this way the carbon it contains is oxidized.

Sometimes to complete the oxidation, after ordinary acid treatment the material is dried and treated with perchloric acid. Eventually it should dissolve completely but sometimes it does not. Or even if it does, the element being tested for may still be combined with organic materials and as a result improper values are obtained when standard methods of analysis are applied. Thus, for example, when muscle is tested for chloride by the commonly used Van Slyke method, the muscle is dissolved by heating it in concentrated nitric acid, and at the same time the chloride is precipitated by silver nitrate. Titration for chloride is then done by the standard Volhard method. Actually, this procedure, though it has been widely used, gives incorrect values for chloride.

If the muscle tissue is really ashed by ignition in a tightly sealed container such as a Parr bomb, values for chloride can be obtained which are approximately twice those that can be obtained by the Van Slyke method. Even with the spectrographic methods, difficulties in ashing may be encountered. In the flame photometer the material is used in solution. A small sample of the solution is then ignited. But in order to obtain a representative sample, it is necessary to have either a clear solution or an even suspension. Sometimes in preparing material for analysis, a few relatively large flakes of precipitate may appear.

This can be very annoying, and can prevent proper analysis. Similar difficulties may arise in ordinary spectrography. Thus, in interpreting the results reported in the literature, it is often necessary to view these cautiously. In the discussion that follows, no attempt will be made at completeness. Over 34 elements are commonly found in plant and animal tissues. Of these, carbon, hydrogen, oxygen and nitrogen are constituents of such organic compounds as proteins, fats and carbohydrates; also, to a less extent, sulfur and phosphorus. Other elements may likewise be combined with the proteins of the cell, but ordinarily they do not form an intimate part of the protein molecule. In addition to the elements, the following are among the elements found in practically all types of living substance: potassium, sodium, calcium, magnesium and chlorine. However, many plants can grow in the apparent absence of sodium and, indeed, some of the common nutrient solutions used for the growth of plants contain no sodium.

Lithium is typically present in marine animals, but according to Fox and Ramage, lithium, although widespread, is not a universal constituent of protoplasm. Rubidium is also rather widespread in its occurrence. It was found in all human tissues examined except bone, being especially abundant in heart and striated muscle. Although higher plants seem to require calcium, some fungi are said to grow normally without it. It is by no means an easy matter to remove traces of sodium or calcium from nutrient solutions, so that it is difficult to determine whether or not living material can exist in the total absence of these elements. Strontium is close to calcium chemically and appears to be able to replace it in some biological systems. A small amount of strontium is generally present in marine invertebrates. Magnesium is a constituent of chlorophyll, and is therefore found in all green plants. Although sea water itself contains several times as much magnesium as calcium, the animals which live in sea water commonly contain less magnesium than calcium. This is not always true, however, and some gastropods including the very common marine snail Littorina littorea contain more magnesium than calcium. Some insects are also richer in magnesium than in calcium. Rats cannot live without a small amount of magnesium. If deprived of it, the calcium-magnesium balance is upset, and convulsions and death follow. Boron appears to be an essential constituent of citrus plants, tobacco, and tomato. It is also found in small amounts in Marine Invertebrates. Aluminum is a very common constituent of living material, especially in plants. It is also found in animal tissues. Heavy metals, although not present in large amounts, are very important for life. In organic combination they act as carriers of oxygen (in the blood) and as oxidizing catalysts within cells. Especially important are iron and copper.

Iron is a necessary constituent of the blood cells of all vertebrates and of many invertebrates. As Warburg first showed, cellular respiration is primarily dependent on the presence of iron-containing catalysts believed to be present in all living cells. The blood cells of many invertebrates contain copper and the blue colour of crab and limulus blood is due to the presence of this element. Lindow, Elvehjem, and Peterson have shown a widespread distribution of copper in living tissues; Sheldon and Ramage conclude that it is a universal constituent of living protoplasm.

This opinion is shared by Marston. The literature on the copper content of living systems is now very large, mainly because of the importance of copper in cell metabolism. Zinc is widely distributed among both plants and animals. Galtsoff found it in many marine invertebrates. It has also been reported for the silkworm and for vertebrates generally. Gettler and Bastian describe a simple method of detecting zinc in tissues and they find rather high amounts (50 mg. per kilo) in human liver and kidney. Keilin and Mann showed that zinc is a constituent of the important enzyme carbonic anhydrase and Smirnov finds that the zinc content of the red blood cells of various vertebrates runs parallel to their carbonic anhydrase content. Zinc is also present in blood plasma. The retina, iris, and choroid of vertebrate eyes and especially the eyes of fish are rich in zinc and this is also to be correlated with a high carbonic anhydrase content of these tissues. In view of the fact that the insulin molecule is thought to contain zinc, it is not surprising that the pancreas should be rich in this element. Manganese is also an important element.

It is essential for the growth of rats and chickens, and also for the growth of various types of plants. Webb found manganese generally present in invertebrates; according to Vinogradoff, ants have an especially high content. Human blood contains only a small amount of it. In almost all plant and animal tissues, it is possible to detect cobalt and nickel by means of extremely delicate tests. The quantities found range from 0.002 to 2 mg. per kilo. All types of animals that have been tested contain molybdenum; tunicates are especially rich in it. Marston (loc. cit). believes that molybdenum is probably always present in living matter. Some metallic elements are of rather sporadic occurrence. Small amounts of arsenic and of silicon have been found in various tissues of man and higher animals.

Ohlmeyer and Olpp found quite appreciable amounts of silicon in the blood, lung and pancreas of the calf. It had previously been found in connective tissue, hair and feathers. Some plants seem to have a definite need for silicon. Webb found tin in the starfish, the sea urchin, and in a gastropod mollusk. There is a literature on the presence of vanadium in thc blood cells of ascidians. The element is also present in other tissues of ascidians but not in detectible amounts in rat tissues. Vanadium occurs in a wide variety of plants. Traces of uranium have been found in the fungi Aspergillus niger and Ustilago carbo; also in human kidney, brain, thyroid and blood. The rare element gallium is apparently essential for Aspergillus. Many human organs contain mercury in small amounts. In some parts of the world gold is found in the soil and is taken up by plants. Deer eating such plants accumulate gold in their horns, hoofs and hair. There is a large body of evidence to show that metallic elements present in traces have a profound effect on the life of various types of plants. The subject is of great practical importance. Of the non-metallic elements, chlorine is probably found in all types of living material.

For many years, it was believed that chlorine although present in tissues generally, did not occur within the cells but only in the spaces between the cells. This doctrine was at one time held for various types of vertebrate tissues and organs, but especially for muscle. Analysis of muscle by the standard Van Slyke method showed relatively low concentrations of chlorine, and this could all be accounted for if it were assumed that the tissue spaces of the muscle contained chlorine in the same concentration as that in which it occurs in the blood.

As a matter of fact, as already noted, the Van Slyke method gives improper values for the chloride of muscle. Moreover, chloride in reasonably high concentrations can actually be demonstrated in isolated frog muscle fibres. Fluorine is of little importance for animals and plants, for with few exceptions it occurs only in minute amounts and it is not known to have any well-defined function. The same applies to bromine. As is commonly known, iodine is an important constituent of the thyroid gland of vertebrates. Iodine is also abundant in marine brown algae and is obtained commercially from this source. This analysis differs from most analyses of muscle in that it shows more sodium than potassium. Typically muscle is much richer in potassium than in sodium. It will be noted that the sum of the cation equivalent weights is decidedly greater than that of the anions.

Organic Constituents

Living substance contains three main types of organic constituents. Of these the proteins and carbohydrates are typically water-soluble substances. It is true that some proteins and some carbohydrates are not soluble in water or in dilute salt solutions, but insoluble proteins are relatively rare, and insoluble carbohydrates are easily broken down to soluble compounds. On the other hand, the third class of substances characteristic of protoplasm is distinguished by a thorough-going insolubility in water.

Indeed, all protoplasmic constituents not soluble in water, but soluble in fat solvents, are grouped together as lipids. The term "lipids" can then be taken to include, first, the true fats, secondly the phosphatids which are compounds of fatty acid, phosphoric acid, and organic base, and, thirdly, the sterols. The latter bear no chemical kinship to the fats, but resemble them in being insoluble in water but soluble in ether. Although not so abundant in living material as are the proteins, the lipids are indispensable for life.

Insoluble in water, they are nevertheless finely dispersed in the watery protoplasm. This is in part due to the fact that the lipids are emulsified; in part, also, to the fact that the lipids may combine with proteins to form water-soluble compounds. The nature of these compounds is not at all well known, but it seems certain, as already noted, that they do occur in protoplasm. The combination of lipid with protein is but one example of the fact that within the living cell various types of chemical compounds may be linked to each other. At the present time, our knowledge of actual combinations of organic substances in the living protoplasm is meager. What we know chiefly is that the organic substances which do constitute the protoplasm are of such a nature that they might well be assumed to combine.

Thus, if some of the proteins contained in cells behave primarily like acids and others like bases, it is only natural to assume that these proteins might be linked together in life. Indeed the existence of combined proteins is well known. Carbohydrates may be combined with proteins not only in true chemical combination but also the carbohydrates may be tied to the proteins by loose secondary linkages. Various alkaline and alkaline earth cations can be and undoubtedly are bound to the proteins of the protoplasm, and they may also be linked to lipids or even to carbohydrates. Perhaps also anions may be bound to the proteins of cells. Indeed, although no one at the present favours the old biogen concept of protoplasm, there is a growing fund of knowledge to indicate that many of the chemical compounds isolated from dead cells may in the living organism be linked by chemical bonds. In view of the fact that living cells are surrounded by osmotic membranes, and are thus tiny osmotic engines, it is important for the physiologist and for the general biologist to know something about osmotic phenomena and the laws of osmosis.

Indeed, in any investigation in which cells or tissues are immersed in a solution it is essential that this solution have the same osmotic properties as the fluids in which the cells and tissues are normally bathed. Similarly, if a solution is injected into the blood stream of an animal, it should not differ markedly in osmotic properties from the blood itself. Many a biological investigation has been spoiled by a failure to observe this rule, or by an ignorance of how to arrive at a proper osmotic concentration. In the following discussion no attempt will be made at completeness, and very little attention will be paid to theories of osmosis or to mathematical treatment of osmotic laws.

Moreover, although solutions in all types of solvents show osmotic behaviour, we shall confine our attention to watery solutions, for these are the only types of solutions with which we shall have to deal in living systems. When two solutions are separated by a semipermeable membrane, there is a flow of solvent towards the solution of higher molecular concentration. This constitutes osmotic flow, and the pressure developed as a result of this flow is osmotic pressure.

A semipermeable membrane is a membrane which permits solvent to pass, but not dissolved substance or solute. Actually no perfect semipermeable membrane is known; at any rate, we know of no membrane which permits water to pass and resists the passage of all dissolved substances. However, there are membranes which are permeable to water and impermeable to certain dissolved substances, and if we choose the proper solutions, these membranes are truly semipermeable.

ANIMALS AWARENESS

Behaviourism claims that consciousness is neither a

definite nor a usable concept. Its closest scientific companion is physiology. It is different from physiology only in the grouping of its problems, not in fundamental or in central viewpoint." Biologists concerned with animal behaviour have adhered, with very few exceptions, to the comparable tenets of Jacques Loeb. Loeb was confident that virtually all behaviour of animals and men could be analysed successfully in terms of tropisms, or forced movements.

Morgan's canon ("In no case may we interpret an action as the outcome of the exercise of a higher psychical faculty, if it can be interpreted as the outcome of the exercise of one which stands lower in the psychological scale") has been widely interpreted as requiring that complex functions should not be postulated if a simpler explanation will suffice. This is, of course, the widely accepted principle of parsimony; given a choice of two or more plausible explanations, the simplest is preferred. Similar reasons of parsimony are usually given against any suggestion that animals might have mental experiences, or be aware of the flow of events with which their behaviour interacts.

As a result, only the most physiological explanations have customarily been recognized as worthy of scientific consideration. In allowing ourselves to entertain the notion that animals may be aware of past, present, and future events by means of mental images it certainly is not necessary to assume that such mental experiences are at all similar to those which a person might have under analogous circumstances. This uncertainty has been one of the chief reasons why many behaviourists have denied that it is meaningful to inquire about the properties of human mental experiences, let alone any that animals might have. But ethology has now developed to a point at which these behaviouristic inhibitions may well have become a barrier to further progress, because they discourage recognition of significant opportunities for new and promising lines of investigation.

Almost all linguists, and most philosophers who have considered the question, have vacillated between denying to animals any significant mental experiences and grudgingly admitting the likelihood of certain simple ones, while rejecting others. The latter are then held to be of crucial importance. These discussions are often eloquent, but they show signs of what ethologists call conflict behaviour. To some, this will appear to reflect a fundamental difference between scientists and humanists, and suggest that communication behaviour presents a magnificent opportunity for fruitful interaction and cross-fertilization between broad-minded scientists and equally perceptive humanists. The implicit denial of mental experiences to animals has almost become an act of faith, supported by arguments and assertions that true language is a unique and characteristic attribute of our species.

Most people are determined to hold the line against animals. Grant them the ability to make linguistic reference and they will be putting in a claim for minds and souls. The whole phyletic scale will come trooping into Heaven demanding immortality for every tadpole and hippopotamus. Better be firm now and make it clear that man alone can use language and make reference. There is a qualitative difference of mentality separating us from the animals." Later Brown-recognizes that "If vocalization is acknowledged to be unimportant, the dances of the bee appear to be very much like referential language." But he places great emphasis on the assumption that "the dances are unlike language in that they are not learned." He feels that animal communication is rigid, always predictable when the circumstances are specified, whereas human speech is not. He says of the dancing bees "the followers' reaction is *too* reliable." The impression of mechanical predictability of the responses of follower bees stems in part from the eloquent simplicity of von Frisch's descriptions of this behaviour in his semipopular books and articles.

More detailed reading of his technical papers, or actual observation of the bees themselves, show considerable variability. Many followers do other things instead of flying out to the place indicated by the dances; they may seem to ignore the dances altogether and turn to other activities. Even those that do leave the hive do not all reach the indicated goal.

It is technically so difficult to observe individual bees known to have followed a given dance after they are flying in the open air, that we know almost nothing about their behaviour between the time they leave the hive and the moment they arrive at the feeding place. As pointed out by J. L. Gould bees newly recruited by dances often take much longer to reach the food than the time necessary for a direct flight. Contrary to Brown's opinion, and that of many others, there is quite enough variability in the behavioural responses of bees and other communicating animals to leave room for the assumption of spontaneity. Examples of "philosophical conflict behaviour" are evident when discuss the problems of analyzing mental processes from their respective viewpoints as philosopher, physicist, theologian, and biologist.

When they consider the possibility that nonhuman minds might exist, they reflect the current climate of opinion by devoting much more attention to computers than to animals. At one point, Washoe and other chimpanzees are denied true minds on the ground that they merely mimic the sign language of the deaf, but elsewhere because they have been taught this language by human trainers. However, animals are conceded to have subjective feelings and perceptions, that is, sense-consciousness. Waddington argues that "if consciousness were to be adopted as a criterion of mind, it would be a signally useless one, because the only way to tell whether any other thing is conscious is to ask it. And that we can only do to human beings; the concept of consciousness is not applicable to anything but a language using animal."

Yet some animals, such as dogs and cats, are recognized as capable of having intentions, and Longuet-Higgins admits that: "An organism which can have intentions we think is one which could be said to possess a mind [provided it has] the ability to form a plan, and make a decision—to adopt the plan." From this divergence of opinions, the presence of mental images and an ability to provide introspective reports on self-awareness and intentions emerge as a criterion of mind. Hampshire clearly expressed the opinion of many philosophers: "It would be senseless to attribute to an animal a memory that distinguished the order of events in the past, and it would be senseless to attribute to it an expectation of an order of events in the future. It does not have the concepts of order, or any concepts at all. An intention involves, among other things, a definite and expressible expectation of an order of events in the future, and is possible only in a being who is capable of at least the rudiments of conceptual thought." Yet chimpanzees and other animals have been demonstrated to have at least moderately complex concepts. One example is provided by the laboratory experiments of Rohles and Devine on the "middleness" concept. Chimpanzees were trained to select from a row of objects the one in the middle position, in the sense that equal numbers of objects were to its left and right.

If the objects were arranged symmetrically, chimpanzees could select the middle one when as many as 17 were present. If irregularly and asymmetrically arranged—that is, with irregular spacings between adjacent objects—chimpanzees could solve the problem when up to 11 were presented. This is roughly equivalent to the ability of four- to six-year-old children to solve comparable problems. Rohles and Devine did not succeed in training rhesus monkeys to select middle objects in sets larger than five. The extensive field studies of ethologists have also cast grave doubt on negative assertions of Hampshire and many others.

For example, ethologists have found the term "intention movement" widely applicable to those postures and relatively slight movements of animals that convey reliable information about their probable future behaviour to other animals. There is no doubt from the reactions of conspecifics that such information is indeed conveyed. But it has been the curious custoi of most ethologists to stop short of interpreting an intention movement as evidence that the animal has a conscious intention, although Daanje did state that animals intend to perform particular behaviour patterns. Since both conspecifics and human observers can predict the future behaviour of an animal from its intention movements, it seems remarkably unparsimonious to assume that the animal executing the intention movement cannot anticipate the next steps in its own behaviour. These reductionist attitudes are very resistant to erosion, however, and even the broadly perceptive biological commentator Lewis Thomas writes: "A solitary ant, afield, cannot be considered to have much of anything on his mind; indeed, with only a few neurons strung together by fibres, he can't be imagined to have a mind at all, much less a thought.

He is more like a ganglion on legs." Is the dancing honeybee a ganglion on wings? Bennett argued with charming erudition that to qualify as rational creatures bees would have to exchange more abstract messages than real bees had been demonstrated to do. But Bennett's ideas about real bees were apparently based on von Frisch's first reports of his discoveries, and Bennett did not consider the implications of the later discovery that the dances are used to communicate about things other than food or to exchange information about hive sites. As discussed by J. L. Gould many of Bennett's criteria for rationality, which he asserts are not satisfied by the dancing bees, have lost their validity, either because recent observations have shown that bee dances do satisfy the criterion in question, or because they have never been looked for by observations or experiments that could reveal either their presence or absence.

As pointed out above in connection with the views of Brown, animal responses to communication signals are, in fact, quite variable. In this connection, it is commonly stated that no animal can use its communication system to tell a lie. Of course, a lie requires intention to deceive, so that to judge whether variability in animal communication behaviour is "noise" or prevarication requires knowledge of the animal's intentions. But conscious intention is a category of mental experience that is widely believed to be uniquely human. Maritain exemplifies this climate of opinion: "Animals possess a variety of means of communication but no genuine language no animal knows the relation of signification or uses signs as involving and manifesting an awareness of this relation in the last analysis the relation of signification remains unknown to the bees. They use signs—and they do not know that there are signs. The whole thing belongs to the realm of conditioned reflexes, whereas language pertains to the realm of the intellect, with its concepts and universal notions." Grice states the generally accepted psychological conditions of language more rigorously: "Perhaps we may sum up what is necessary for A to mean something by x as follows. A must intend to induce by x a belief in an audience, and he must also intend his utterance to be recognized as so intended." There is a serious circularity in these arguments: conscious intention is ruled out *a priori* and then its absence taken as evidence that animal communication is fundamentally different from human language. It seems that the more difficult the question under consideration and the less adequate the available evidence, the more definite become the generally accepted assertions about the differences between human language and animal communication.

But it is important to ask on what basis such definite assertions can be made about what bees and other animals do or do not know. Have we allowed nonscientific value-judgments to colour our thinking about these questions? For example, Adler concludes that the communication system of honeybees is "a purely instinctive performance on their part and does not represent, *even in the slightest degree,* the same kind of highly variable, acquired or learned, and deliberately or intentionally exercised linguistic performance that is to be found in human speech."

He then goes on to argue that if it were to be established by some future investigations that animals differ from men only in degree and not radically in kind, we would then no longer have any moral basis for treating them differently from men, and, conversely, that this knowledge would destroy our moral basis for holding that all men have basic rights and an individual dignity that render it wrong to mistreat groups of men judged to be inferior for the benefit of supposedly superior groups.

Followed to its logical conclusion, this argument implies that the comparative investigation of communication behaviour has more dangerous potential consequences than nuclear physics had in the 1930s, or the current fear that synthesis of certain new forms of DNA might produce uncontrolled pathogens. The narrowing gap between animal communication and human language calls into question not only the assumption of a qualitative dichotomy, but also the accompanying assumption that animals respond mechanically to external or internal stimulation, whereas human beings speak with conscious understanding and intent. Despite the deep reflection and thoughtful eloquence of philosophers, the nature, and even the existence, of mental phenomena have remained largely outside the scope of natural science. Occam's razor and Morgan's canon have been so seriously adhered to since the 1920s that behavioural scientists have grown highly uncomfortable at the very thought of mental states or subjective qualities in animals.

When they intrude on our scientific discourse, many of us feel sheepish, and when we find ourselves using such words as fear, pain, pleasure, or the like we tend to shield our reductionist egos behind a respectability blanket of quotation marks. There have been a few exceptions to the behaviouristic tradition and the related stress on parsimonious explanations in biology. Adams discussed what he felt were logical weaknesses in Morgan's original statement of his widely cited canon, which he found "insufficient to criticize the inference of mind." One problem with Morgan's canon is that it is based on an intuitive classification of behaviour into higher and lower categories, the latter to be preferred unless the evidence forces postulation of the former.

But no definitive, objective reasons are provided for assigning behaviour to particular places on the scale of lower to higher. Adams advocated that mental experiences could reasonably be inferred in another animal to the degree that its "structure, situation, history, and behaviour" resemble those which accompany such mental experiences in a human observer. Jennings repeatedly and eloquently argued the case for an open mind concerning behavioural and mental continuity between men and other animals. Marler believes that "we delude ourselves if we think that a complete [behavioural] discontinuity separates us from other animals." And Boyle advocates that "the psychologist will, if he is wise, acknowledge that organisms make sense of their experiences, and he must attempt to discover what this sense is. Unfortunately psychology has turned its back on this task because, since a psychologist's hypotheses about an animal's experience cannot be confirmed, psychologists have to a large extent ceased trying to understand other human beings as well."

Yet Boyle, although ready to admit that when a cat rubs against his feet she intends to induce him to give her food, nevertheless finds it "difficult to imagine that a bee *intends* to communicate to fellow workers a message about the distance and direction of pollen. That may be the meaning of the dance to an observer, but it is doubtful whether bees are capable of this type of understanding." Small size or phylogenetic remoteness from man are evidently taken as evidence against any form of conscious intent in bees. The image of a "ganglion on legs" dominates our view of invertebrate animals. Tolman developed a "purposive behaviourism," in which what are ordinarily considered as mental events and processes were treated as intervening variables between external stimuli or internal influences on the one hand and overt, observable behaviour on the other. By simply designating mental processes as intervening variables, the difficult question of their nature and reality is avoided. But Tolman's position is less rigid than strict behaviourism, and he accepts the reality of conscious awareness in animals, for example in a white rat at the moment of learning some new behaviour pattern, such as a specific portion of a maze. In recent decades, however, most psychologists have either avoided this question or taken positions closer to strict behaviourism.

Many schools of philosophy dissent vigorously from materialism and from logical positivism, and not only do these schools accept the reality of concepts which behaviourists reject as meaningless; they often attach central importance to them. But philosophers of this kind seldom pay much attention to animals. There is an active discipline of cognitive psychology which feels free to deal with *human* mental experiences, although often refraining from the use of explicit mentalistic terms. Neisser asserts emphatically that "Cognitive processes surely exist, so that it can hardly be unscientific to study them." Schultz reviews the trend for some contemporary psychologists to abandon the strict taboos of behaviourism. When Irwin considers many of these broad and challenging questions within a fairly conservative framework of objective analysis, he does recognize that men and animals have "expectancies." By this term, Irwin seems to mean something closely resembling mental or internal images of the possible future outcomes of various alternative patterns of behaviour. The term consciousness, however, is still one which Irwin struggles to avoid, but Kimble and Perlmuter go so far as to speak of volition.

Fodor, Bever, and Garrett review the contributions of psycholinguists who have followed the lead of Chomsky in rejecting a behaviouristic position regarding *human* language, and find it essential to consider mental entities. But few psychologists or even ethologists have yet moved away from an essentially behaviouristic position with regard to animal behaviour. Lorenz is a notable exception among ethologists; he does not hesitate to express a belief that animals have subjective experiences, although, like Adams, he concentrates his attention on higher vertebrates.

Bertrand hesitantly ventures to speak of a monkey's behaviour as voluntary. Razran briefly considers the possibility that animals have simple thoughts. Brewer has recently argued in the course of a vigorous dissent from behaviourism: "Since cognitive theory holds for humans, it is unparsimonious not to apply it to animals." And Weiner states that "cognitivists also use the man-animal continuity to promote their view that even infrahuman behaviour is guided by cognitive processes."

Mowrer struggled to escape the rigid restrictions of the behaviouristic position while still dealing only with observable events, but was nevertheless led to state that "if consciousness were not itself experienced, we would have to invent some such equivalent construct to take its place." Yet he concludes that bee dances are "limited to 'sentences' of the thing-sign variety" because samples of food are transferred, rather than being represented by a symbol. An animal's anticipation into the future, and he may well be correct in this estimate for a dog in the situation suggested. But suppose a dog *did* anticipate events likely to occur tomorrow. How could we recognize this fact from observing its behaviour today? Lacking any evidence at all, we make a negative judgment, reasonably enough, but tend to forget its weak foundation. Miller asserted that "Man is the only animal to have a combinatorily productive language a species-specific form of behaviour Serious attempts have been made to teach animals to speak. These attempts have uniformly failed in the past and, if the argument here is correct, they will always fall in the future."

Recent successes in teaching chimpanzees to communicate with combinations of gestures clearly cast doubt on the validity of Miller's "pre-Washoe" prediction. Rensch and Popper expressed similar views. Miller, Galanter, and Pribram Langer and others state without qualification that man is the only animal that can be aware of his own future death. But we suggest that we pause and ask just how does anyone know this? What sort of evidence is available either pro or con? Suggestive inferences can be based on the clear demonstration that many social animals recognize each other as individuals, and on the observation that some animal mothers show signs of distress over the corpses of their dead infants, which they carry about for days.

Eccles has proposed an explanatory framework for brain function that includes conscious experience, which he feels exists only in the human dominant cerebral hemisphere. He cautiously reserves judgment concerning the existence of consciousness in the subordinate hemisphere of the human brain and in the Great Apes. Sperry goes farther, and not only recognizes the importance of conscious experience in the human brain, but concludes that it "exerts a directive holistic form of control over the flow pattern of cerebral excitation." Much of the evidence on which Eccles and Sperry have based these ideas comes from surgical cutting of the corpus callosum in human patients suffering from severe epilepsy.

When the hemispheres are thus deprived of their normal channel for exchanging information, they can learn to recognize quite separate sensory stimuli, and they seem to operate almost independently with respect to complex behaviour. This is the case in both human patients and all other mammals tested so far. In some patients, only the dominant hemisphere can report verbally about its sensations and learned behaviour. But in others, mental functions seem more evenly divided between the hemispheres. In some of the patients with a severed corpus callosum, the subordinate hemisphere can mediate learning to recognize objects, and the patient can demonstrate that learning has occurred by pointing to the correct object when asked to match a sample. Yet the same subject is unable to report the correct choice orally or in writing.

It seems clear that the subordinate hemisphere carries out many of the mental functions ordinarily considered conscious, but lacks the ability to report them in words. As Eccles puts it, "the minor hemisphere resembles an animal brain." These discoveries add to the evidence for physiological continuity between men and animals in brain function, and suggest a comparable continuity in mental experiences. Pribram and Gazzaniga review much the same evidence and reach conclusions similar to those of Eccles, although they are expressed in slightly different terms.

ORIENTATION AND COMMUNICATION

Many behavioural scientists express feelings of discomfort, or even outrage, at the inference of conscious intention in animals because previously unsuspected complexities in their orientation and communication have been discovered. On strictly logical grounds, complexity of behaviour and conscious awareness are neither commensurate with or necessarily related to one another in any way. Inanimate mechanisms can be enormously complex and difficult to understand, but most *descriptions* of animal behaviour can be modeled by mechanisms far simpler than a television receiver.

The same can be said of many physiological mechanisms. For instance, a very simple electronic circuit can produce an electrical signal closely resembling the spike potential of a neuron. But only the most naive engineer-turned-neurophysiologist would accept the existence of such a circuit as a satisfactory explanation of the functioning of nervous systems. To Loeb, the existence of a phototactic machine constructed out of wheels, electric motors, and photocells was evidence for believing that animal, and even human, behaviour could be explained in terms of tropisms.

But in the 1970s, the crippling limitations of such intellectual mypoia should be clearly apparent; the simplicity often lies not in the behaviour, but in its description. Can we accept the reality of our own conscious awareness but reject the hypothesis that any pattern-recognizing machine is also consciously aware of the pattern to which it responds selectively? It certainly seems easier to reject the notion of consciousness in a simple mechanism, such as a lock built to accept a particular key, than in a computer system programed to respond correctly to an especially intricate input signal.

This is primarily because we can understand how the lock works, but find it impossible to encompass in our own mental imagery the complexities of the computer Programme. Must we therefore infer conscious awareness whenever we do not know how a mechanism works? Complexity of some kinds surely provides no convincing evidence for the existence of mental experiences. The physiological mechanisms by which a kidney regulates the chemical composition of the blood, or the biochemical systems that regulate cellular respiration, are marvelously complex and incompletely understood, but certainly are far different from central nervous systems in their structure and function.

Analogies can be drawn by describing all three in terms so general that they apply to any self-regulating system, and these system properties are of interest in their own right. But brains and minds, insofar as we allow ourselves to admit the existence of the latter, are surely different from kidneys and mitochondria in basically significant attributes. In the interpretation of communication behaviour, pride of parsimony can lead us into some awkward situations. The waggle dances of honeybees convey to other bees not information about distances and direction, or about the actual location of a distant food source, but only information about the internal state of the dancer; that is: "A forager bee does not state that it has found food at a certain place; rather, she describes a direction of flight (perhaps a flight she is likely to make again shortly) and, on request, provides a sample of the food."

But suppose we knew nothing about human language and watched and listened to human beings while they were conversing. We might well conclude that they, too, were simply describing their internal states. Indeed, the definition of internal states can easily be extended to cover the most complex speech or writing, which can, if one wishes, be interpreted as "merely describing the internal state" of speaker or writer. Pushed to their limits, all these arguments make sense only on the implicit assumption of conscious intent on the part of human beings and its absence in animals.

The general feeling that our species is uniquely superior has suffered a series of intellectual setbacks beginning with the Copernican and Darwinian revolutions, the second having far more basic repercussions than the first. Later, the ability to learn from past experience was advanced as a unique human attribute, but successive discoveries of learning in animals more and more distantly related to us forced abandonment of that criterion. Tool-using suffered a similar fate, for example from studies of Darwin's finches and sea otters, as did tool-making more recently at the hands of chimpanzees. When evidence is presented that a nonhuman species achieves some of the previously proposed criteria for distinguishing human language, "the list grows longer in order to exclude the interloper species. If this kind of progression continues we may eventually have a definition of language that isomorphically maps the behaviour of human language; in essence a redundant description of the behaviour". Similar viewpoints have been presented in semipopular fashion by Linden and S.

J. Gould. Most biologists concerned with animal behaviour seem to share the basic views of the linguists and philosophers quoted above—for example, Simpson and Dobzhansky. When ethologists discuss animal communication, we almost never find any suggestion that subjective experiences may occur in the animals under study. The reductionist, behaviourist tradition still dominates our thinking. When von Frisch or Lindauer find that bees dance in patterns correlated with the location of food or something else needed by the colony, or that a bee which has danced about one potential location for the swarm shifts to a different site under the influence of the more enthusiastic dances of others, they do not suggest that perhaps the bees feel any need for sugar, pollen, water, or a suitable new cavity for the colony. That is taboo.

Virtually all descriptions and discussions are in terms that would be equally applicable to a living animal or an appropriately contrived machine. This viewpoint has served our science well for more than 50 years by constraining speculations and focusing attention on phenomena amenable to experimental analysis. Seventy years ago, this disciplined restraint was a healthy reaction to an earlier tendency to ascribe human feelings to a wide variety of animals, mostly on the basis of rather unconvincing anecdotal evidence. But suppose the biologists and psychologists active at the turn of the century had known about the communication dances of bees and about the recent insights the Gardners have obtained from Washoe.

Would they have been so adamant in banishing from respectable consideration all notions of mental states in animals? Their message was: As a working strategy of research, assume no mental states or subjective experiences, and see how much of animal behaviour can be accounted for on this parsimonious basis. This has now been done on a large scale, and some of the results suggest that it is time to review our perspectives and strategies in the light of the new discoveries. The importance of this question is demonstrated by the heavy reliance of linguists and philosophers on the consciously intentional use of language as the principal distinguishing characteristic of our species. A major reason for these philosophical assertions has been the acceptance by those linguists and philosophers of the general conclusions expressed by students of animal behaviour. It is suggested that behavioural scientists now have the opportunity, and perhaps an obligation, to explore and discuss the implications for this traditional, behaviouristic viewpoint of recent discoveries about communication behaviour in animals. When the question of subjective feelings or conscious intent in animals is raised, the customary response of many biologists or psychologists is to take the cautiously agnostic position that subjective or mental states, together with consciousness and intention, are beyond the reach of objective scientific inquiry.

The virtual extinction of introspective psychology, under the onslaughts of behaviourism in the 1920s, seemed to be a firmly closed story. Yet, when the behaviouristic position is stated at its scholarly best—for example, by Lashley it is essentially agnostic. It does not deny the existence of mental states, but argues that they are one and the same as neurophysiological processes, and that it is unprofitable to attempt any sort of scientific analysis based on introspective reports. Half a century of behavioural science has progressed on this basis, along with many discoveries in neurobiology in the broadest sense, including ethology. There remains that tendency for what was originally an agnostic position to drift implicitly into a sort of *de facto* denial that mental states or consciousness exist outside our own species. It is very easy for scientists to slip into the passive assumption that phenomena with which their customary methods cannot deal effectively are unimportant or even nonexistent.

To quote Fouts: "All one needs to do is to look around and *not* see something and then conclude that the thing that was not seen in a particular species is totally absent in that species." Here we should also like to follow the example of Holloway in quoting Daniel Yankelovich: "The first step is to measure whatever can be easily measured. This is okay as far as it goes. The second step is to disregard that which can't be measured or give it an arbitrary quantitative value. This is artificial and misleading. The third step is to presume that what can't be measured easily isn't very important. This is blindness. The fourth step is to say what can't be easily measured really doesn't exist. This is suicide." Biological evolution is universally accepted by behavioural scientists as a historical fact. Animals are used as surrogates or "models" for behavioural investigations on the implicit assumption that principles discovered in this way are applicable to our own species, as has been abundantly confirmed in physiology, biochemistry, and medicine. Certainly this assumption implies qualitative continuity. If, for example, human learning were believed to be radically different in kind from that available for analysis in other animals, no one would even suggest applying to questions of human education what has been learned by studying rats, pigeons, or monkeys. Yet, when questions of communication and language arise, even hard-nosed behaviourists take for granted a large element of discontinuity.

To argue that language is unique to man and, therefore, no matter how complex animal communication turns out to be, it cannot possibly be continuous with human language, is indefensibly circular. Must we reject evolutionary continuity in order to preserve our gut feeling of human superiority, as proposed by Adler, Critchley, and Langer. Or can we be satisfied with a merely quantitative, if enormous, difference between communication behaviour in our own and other species? If we insist on a qualitative human-animal distinction in the area of communication behaviour, a radical difference in kind in Adler's terms, must we support our insistence by criteria as subjective and difficult to test as those that were rejected by the founders of behaviourism? The rigid position of the strict behaviourists is now being questioned with increasing frequency.

Many of us were taught, under pain of banishment from professional psychology, never to use these terms, at least not during 'working hours' such language was deemed completely unsuited to the purposes of science. But it is slightly ironical that those very methods of analysis and research which radical

Behaviourism introduced are now leading us, ineluctably, back to concepts which Behaviourism was determined to ignore—or even destroy." plausible one. It is also instructive to ask where one should draw a line to represent the boundary of scientific validity. Very strict behaviourists might stop after Affect, others may venture farther down the list. There are, of course, many philosophers who disagree with positivism, and they feel comfortable with a list extending beyond this one in the direction labeled Taboo. Perhaps Jennings and Thorpe have outlined the most reasonable view, considering the limited evidence available; namely, that the gradient is a true continuum without sharp discontinuities.

Furthermore, it seems more likely than not that certain animals have mental experiences involving, to varying degrees, the attributes represented crudely by this rank-ordered list of terms. Many branches of science have made significant and substantial progress by employing postulated entities that could not be observed directly, at least when first developed, but which were inferred from observations of their supposed causes and effects. Gravitation, electric potentials, magnetic fields, atoms, neutrinos, X-rays, chemical bonds, hormones, genes, nerve impulses, and magnetic monopoles are pertinent examples. The impossibility of neatly verifying the existence of mesons or quarks has not inhibited theoretical physicists. Astrophysics is likewise based on concepts about events and processes immune from direct observation by any methods we can yet imagine. Investigators of behaviour have attempted to formulate comparable explanatory concepts, such as motivation, drives, or Lorenz's specific-action potential. But perhaps we have been overlooking more directly pertinent concepts lying close to hand or even closer—inside our own heads.

When thinking about Washoe in the act of exchanging information about objects, actions, or desires via coded manual gestures, or when contemplating Lindauer's swarming bees dancing about the suitability and location of cavities where the swarm might find a new home, we submit that it may actually clarify our thinking to entertain such thoughts as

"Washoe *hopes* to go out for a romp, and *intends* to influence her human companions to that end," or "This bee *likes* one cavity better than the other, and *wants* her swarm to occupy the preferred one." Of course, the use of such terms as *want* or *like* does not explain the basic cauces of the observed behaviour or of any mental experiences that may accompany it. Nor should the use of these or similar terms be taken to imply identity with any human mental experiences.

The degree of similarity or difference would be an appropriate question for future investigations. Perhaps this return to a consideration of basic subjective qualities can supply a unifying framework into which many complexities of animal behaviour can be fitted. To paraphrase Boyle, perhaps we can understand how, and to what extent, animals make sense of the flow of events of which their behaviour forms a part. Most people not indoctrinated in the behaviourist tradition take it for granted that animals do have sensations, feelings, and intentions.

This intuitive impression is based on our experience with patterns of animal behaviour that appear sufficiently analogous to some of our own behaviour to permit us to emphathize. The dilemma of contemporary behavioural scientists results from our indoctrination that *as scientists* we must put such notions behind us as childish sentimentality unworthy of a rigorous investigator. Yet the behaviouristic and reductionistic parsimony typified by Watson and Loeb may have led us down a sort of blind alley, at the end of which we find ourselves defending to the last, at least by implication, a denial of mental experience to animals, a denial which we cannot justify on any explicit basis except the presumed absence of communication with conscious intent.

As learnedly discussed by Malcolm, even Descartes, the fountainhead of the philosophical view that animals are merely machines, admitted that they could feel pain or pleasure and express passions. Yet many behaviourists believe that, it makes no difference whether one thinks in terms of possible mental experiences or simply in terms of stimuli and responses, however complex. The same argument can be applied to people. Inasmuch as we have only indirect evidence about *their* mental experiences, we may logically question whether these really exist. But if questions are raised by others about the reality of one's own subjective feelings, who is likely to fall back on a negative, or even an agnostic, response?

NURTURE OF MENTAL EXPERIENCES

Neurophysiological processes is to raise the naturenuture question with regard to mental experiences themselves. To the extent that they are dealt with at all by scientists, it seems to be tacitly assumed that mental experiences result solely from individual experience and, in particular, from learning. This implication is clear in the statements of Pollio, Maritain, and Adler, quoted above. If we really accept the behaviourists' axiom of psychoneural identity we should face up to the possibility that a nervous system might attain those properties leading to mental experiences primarily on the basis of genetic information.

The development of mental experiences might depend on environmental influences only in the general and unspecific sense that DNA cannot lead to a complete animal without an environment that provides the necessary nourishment and other conditions. We might therefore conclude that the behaviouristic argument leads to the likelihood of something akin to "innate ideas" in the philosophical sense. The nature-nuture issue with respect to behaviour has aroused some of the most violent passions and heated debates known among scientists. As is usual in such cases, balanced consideration strongly suggests that both sides are partly correct and that individual experience and genetic heritage each has significant effects on behaviour.

It is also obvious that the relative importance of these two major factors varies widely between behaviour patterns and groups of animals. This means that two extreme, all-or-nothing positions are clearly and equally untenable. Furthermore, interactions between genetic and environmental factors are of considerable importance, and this, together with the enormous difficulty of controlled experiments, make it impracticable in most cases to estimate their relative importance with anything approaching adequate accuracy. This does not mean that either can safely be assumed *a priori* to be all-important (or unimportant) in the absence of direct evidence of a sort that is rarely available at present.

Applying the same balanced approach to mental experience leads to a cautiously open mind concerning the possibility that both genetic and environmental influences, and interactions between them, may be important in the causation of mental processes, including awareness. Knowing so little about mental experiences in other species, we can scarely begin to attack the nature-nuture question, despite its potential importance. But we should not overlook the reality of the question any more than it seems sensible to ignore the possible existence of mental expressions in more than one species. Virtually all comparative psychologists—and ethologists, as well—are at least *de facto* behaviourists in the sense that they concern themselves only with observable behaviour and shun any involvement with possible subjective qualities or mental experiences. The strict behaviourist believes that it is operationally meaningless, and hence foolishly unscientific, to consider even human mental experiences, because they cannot be observed directly and because verbal reports about them are inconsistent and impossible to verify by any other means. S. R. Brown has clearly stated this dilemma.

"Human subjectivity is a phenomenon that has both interested and eluded social scientists for some time. Perhaps beginning with Watson, the behaviourists rejected introspection and romantic mentalisms, and they did so for sound scientific reasons: As scientists, they knew no way of dealing with these mental goings-on. The rejection, however, was regarded at the time (except, perhaps, in the case of Watson himself) as temporary, pending the development of instrumentation capable of dealing with what previously had proved elusive. The second generation behaviourist forgot the original reasons for the rejections, remembering only the act of rejection itself; the third generation behaviourist was even at a greater disadvantage, being unaware anything was even forgotten." When strict behaviourists asserted that "The supposedly unique facts of consciousness do not exist" they were rebelling against schools of psychology which held that mental qualities were different in kind from physiological processes and hence could never, in principle, be explained in physicochemical terms. Lashley was attacking what he saw as the subjectivists' belief in a separate psychic world, "a unique mode of existence not definable in objective terms." He viewed introspection as "an example of the pathology of scientific method." But he also conceded that "there can be no valid objection by the behaviourist to the introspective method so long as no claim is made that the method reveals something beside bodily activity." Hebb has recently argued vehemently: "Subjective science? There isn't such a thing. Introspectionism is a dead duck." Yet later in the same article he asserts that "Psychology is about the mind: the central issue, the great mystery, the toughest problem of all."

But it is not clear how a central problem can be solved by ignoring most or all of the available evidence. B. F. Skinner was well aware of the danger that important problems might be ignored merely because they are difficult to study. He attempted to deal with what are generally called mental processes, or thinking, while maintaining a consistent behaviouristic position. This he did by treating thinking as *covert verbal behaviour*, in which the speaker and listener are the same person. This position equates thinking with a sort of talking to oneself, which is almost as unobservable as the mental concepts which Watson and the original behaviourists rejected as unworthy of scientific mention.

But Skinner sees an important difference, in that such covert verbal behaviour can be influenced by prior and subsequent events, especially by reinforcement, and thus, like other behaviour, its properties can be deduced from an analysis of its antecedent causes and subsequent results. Because Skinner's behaviourism is constrained to deal only with input-output relationships and methods of reinforcement, it was possible, by stretching its definitions, to include even unobservable, covert verbalizing; at least, Skinner felt justified in doing so. For those who cannot bring themselves to consider the possibility that animals can have even the simplest of thoughts, it may be more acceptable to follow Skinner's lead, despite the searching criticism and rejection of a behaviouristic treatment of human language by Chomsky.

For instance, Skinner's definition of thinking as covert verbal behaviour can reasonably be extended to include not only the literal exchange of words and sentences, but also the many other forms of communication used by human beings, such as signs, gestures, mathematical symbols, and the whole nexus of nonverbal communication. Lashley anticipated important elements of Skinner's position, including this broadening of its scope: "The relation of any integration to the speech and gestural mechanisms is of prime importance for its 'conscious aspects.' Not only is the single certain evidence of consciousness in another person the existence of consistent, rational expressive movements.

The core of 'conscious' integration is the verbo-gestural coordination." We can now include consistently communicating animals within Lashley's definition, on the basis of new discoveries. Extensions of this approach are summarized in McGuigan and Schoonover. Including nonverbal as well as verbal communication in a behaviouristic definition of thinking, along the lines advocated by Lashley and Skinner, seems intuitively to be even less of an extension of the behaviourist position than the redefinition of thinking as covert exchange of words and sentences. To accept into the covert fold verbal, but not nonverbal, communication would scarcely be parsimonious.

Extending Skinner's concept of covert verbal behaviour to covert communication behaviour internally within the human brain, we may then ask, why not also within the chimpanzee brain? Washoe and the other chimpanzees who have learned to use and combine gestures that serve them much as the words in human speech can be assumed to manipulate covertly whatever communication systems they use overtly. Can birds sing to themselves covertly, and bees dance to themselves when motivation for communication behaviour is present but the overt behaviour is not possible for some reason? Some readers may feel they have been led astray and that something must be wrong, since we have reached the unwelcome conclusion that animals can think.

But where is the error? Is it in the argument that if words can be employed covertly, so can nonverbal means of communication? Or in carrying over the notion of covert communication from people to chimpanzees? Is this notion acceptable in apes, but not in birds or bees, and, if so, where and how is a line to be drawn? One objection that can be anticipated is the assumption that animals always express their motivational states immediately, whereas men can inhibit the expression and yet retain the internal communication behaviour. But this seems highly unlikely as a general rule, for in many cases animals clearly retain the memory of some relationship about which they communicate only when circumstances are appropriate.

One example from honeybees may suffice to make this point. If bees from a colony with a severe need for food have been foraging at a certain food source and dancing actively about its location and quality right up to sundown, they will ordinarily do no dancing during the night, but as soon as morning comes they will fly out to the same source, taking now a very different direction relative to the sun.

One could explain this behaviour by postulating, quite reasonably, that they had learned other cues to the location of the food, such as landmarks. But under some circumstances, such bees can be stimulated to dance during the night if appropriate lighting conditions are provided.

They dance with the waggle runs oriented at angles to gravity intermediate between the directions indicated in the evening and morning, with the difference roughly proportional to the time that has elapsed since sunset. Not only does this demonstrate the existence of an endogenous biological clock and continuous correction for the passage of time; it also suggests that the memory of food location and the motivation to communicate about it remain present in a latent, covert state somewhere within the bee's nervous system.

Covert communication behaviour can thus be postulated and its properties deduced from the effects of prior causes and subsequent results, just as Skinner attempts to do with covert human verbal behaviour considered as an objective definition of thinking. The assumption that animals always respond in a rigid, machinelike fashion to immediate stimuli is widespread and resistant to the influence of extensive evidence that animals often defer responses to stimuli.

This evidence comes both from laboratory experiments, in which animals can be trained to respond not immediately, but after varying intervals, and from ethological observations under natural conditions. In general, animals we consider "higher" can learn to make such responses after longer intervals of delay. The views of Terwilliger exemplify how this tendency to think of animals as Cartesian machines is often based on questionable implicit assumptions. Arguing that language is a uniquely human trait, he dismisses the dances of honeybees as something inferior because of their assumed rigidity: "no bee was ever seen dancing about yesterday's honey, not to mention tomorrow's. Moreover, bees never make mistakes in their dance."

Leaving aside the technical error that the dances concern nectar, rather than honey, Terwilliger was either unaware of, or ignored along with the fact that bees can be stimulated to dance during the middle of the night about a food source they have visited the day before and will almost certainly visit again the next morning. These questions are discussed in more detail by Gould. The views of Mowrer quoted in the previous chapter, indicate that many psychologists are relaxing the rigid strictures of Watson's original behaviourism. An agnostic reservation of judgment is clearly the soundest position at present. But it should be an open-minded agnosticism, which recognizes the possibility that anwers not yet available may be obtained from future investigation of questions about mental experiences in animals.

HOMO SAPIENS

Against the notion that animals have mental experiences,

it is often objected that such thinking is anthropomorphic, that is, that it requires ascribing human thoughts to other species. But consider the alternative hypothesis—that mental experiences, like other attributes of animals and men, exhibit a continuity of variation and are not typologically discrete, all-or-nothing qualities totally restricted to a single species. There is no reason to believe that any mental experiences that animals may have must be identical to our own. Indeed, there is no reason to suppose that they are absolutely identical throughout *Homo sapiens*, for instance between men and women, or children and adults.

As Wittgenstein puts it, "one human being can be a complete enigma to another. We learn this when we come into a strange country with entirely strange traditions; and, what is more, even given a mastery of the country's language. We do not *understand* the people. (And not because of not knowing what they are saying to themselves.) We cannot find our feet with them." It is actually no more anthropomorphic, strictly speaking, to postulate mental experiences in another species than to compare its bony structure, nervous system, or antibodies with our own. Subjective qualities have remained largely untouched by the Darwinian revolution, primarily for lack of effective methods for detecting them reliably in other species, let alone analyzing them by scientific methods. The prevailing view implies that only our species can have any sort of conscious awareness or that, should animals have mental experiences, they must be identical with ours, since there can be no other kind. It is this conceit which is truly anthropomorphic, because it assumes a species monopoly of an important quality.

The attitude resembles, in many ways, the pre-Copernican certainty that the earth must lie at the centre of the universe. When we attempt to imagine what an animal may be thinking in a given situation, we are obliged to describe in words (or perhaps in human gestures or expressions) what we suggest the animal may feel. Critics object that the mere use of human language introduces a crippling error, because the animal does not use words. But this may be no more than a methodological problem that could be overcome by careful procedures, as, for example, in recent experiments with chimpanzees. The objection would be critical if, and only if, human thoughts were absolutely identical with words and never had a form not conveyed precisely by a verbal description, and if no other kind of thought existed. It seems highly unlikely that both these statements are correct, even for our own species.

They imply, for example, that new experiences could never occur in the absence of appropriate words already shared by a group of people speaking the same language. This would be a serious limitation of the creativity held by Chomsky and many others to be such an important attribute of human language. Granting that words express thoughts more or less imperfectly, the residue of properties not absolutely coextensive with words might be supplied by covert nonverbal communication in the sense discussed earlier. This possibility is more difficult to evaluate and, by extending the concept of covert nonverbal communication, one might well be able to include all vague feelings, impulses, emotions, and so forth. But whether this can be done in any convincing fashion is not critical to the present discussion.

It is more likely than not that thoughts or mental experiences in people and animals share important properties without being completely identical. A further objection that may confidently be anticipated is that interspecific communication could occur only between very closely related species, say man and chimpanzee, but not man and dog and still less man and bird, while man-to-bee exchanges of information would be quite unthinkable because of the remoteness of any common ancestors.

This widely held view contains the implicit assumption that mental experiences vary rapidly with branching evolutionary lines of descent. But the mere possibility that mental experiences might exist in animals has been so thoroughly ignored by behavioural scientists that we naturally have no data whatever about how such experiences may vary between species. An equally plausible hypothesis is that, as mental experiences are directly linked to neurophysiological processes—or absolutely identical with them, according to the strict behaviourists—our best evidence by which to compare them across species stems from comparative neurophysiology.

To the extent that basic properties of neurons, synapses, and neuroendocrine mechanisms are similar, we might expect to find comparably similar mental experiences. It is well known that basic neurophysiological functions are very similar indeed in all multicellular animals. On this basis, we might be justified in turning the original argument of the strict behaviourists completely upside down.

Because neurophysiological mechanisms appear to be very similar in men and bees, the mental experiences resulting from their operation must, according to this line of reasoning, be equally similar. If this seems an embarrassing conclusion, we can try to escape from it by postulating that neurophysiology is seriously incomplete, having failed, so far, to locate those functions that differ so widely between taxonomic groups that they generate incomprehensibly divergent mental experiences.

The alternative is to postulate some special form of human uniqueness, not demonstrable objectively—an unparsimonious procedure, to say the least. Opening our minds to consider the possibility that at least some animals may have mental experiences is only a first step, though a crucial one that requires a significant departure from the current *Zeitgeist* of the behavioural sciences.

Having taken this step, one is tempted to plunge at once into the very different process of inferring particular mental experiences in specific animals. This second step is as hazardous as it is seductive. It is so easy to guess about the processes occurring in the brain of another species, on the basis of its observed behaviour, that we can easily forget how many such confident conjectures have come to seem implausible as more has been learned about the natural behaviour of the animals concerned.

Careful ethological analysis of preceding and subsequent behaviour can provide some indications of the possible mental experiences that may exist within a given animal under particular circumstances. But we would be on much firmer ground if we could obtain more direct access to whatever events may be taking place inside the animal's head. The reductionist and behaviouristic reactions which have led to the inhibitions discussed above—inhibitions which it now seems timely to relax to some degree—arose for the excellent reason that unsupported and highly implausible interpretations of animal behaviour had become widespread towards the end of the nineteenth century.

George Miller has pointed out that Morgan's canon was proposed and achieved wide acceptance, not with any intention of avoiding introspection, nor with any doubt that mental experiences exist in both men and animals. Rather, as Miller puts it, "all that Morgan hoped for were a few reasonable rules for playing the anthrop-omorphic game."

The view that it is anthropomorphic to postulate any sort of mental experiences in animals may have resulted from a confusion of scientific caution and parsimony with unscientific feelings of human superiority. Nevertheless, there is no doubt that enthusiastic observers of animals are constantly in danger of interpreting their behaviour in more complex terms than is necessary or correct.

Clever Hans is an outstanding example. It seems inherently reasonable that the ability of a horse to notice when a human experimenter had stopped making small counting movements and was waiting expectantly for the horse to end its tapping, is a simpler and far more convincing interpretation than the conclusion that this or other horses could really carry out complex arithmetical manipulations, such as multiplication and division of numbers written on a blackboard.

Thus, an important dilemma must be faced if we allow ourselves, contrary to the exhortations of strict behaviourists, to postulate, however tentatively, the existence of mental experiences in animals. To recognize that any mental experiences animals may have need not be identical, or even necessarily similar, to those of a man under comparable conditions, opens up a wider range of potential interpretation but, at the same time, makes it more difficult to gather convincing data. For the time being, it must suffice to emphasize that the problem is serious. Cautious treading of a middle ground is clearly called for to avoid both of two obviously fallacious extremes:

- The postulation of complex mental activites (such as horses capable of long-division) when simpler ones are consistent with the observed behaviour of the animal and the observed responses of conspecifics to its communication signals.

2

Animal Anatomy

Animals must be able to sense and respond to the environment in which they live if they are to survive. They need to be able to sense the temperature of their surroundings, for example, so they can avoid the hot sun. They must also be able to identify food and escape predators. The various systems and organs in the body must also be linked so they work together. For example, once a predator has identified suitable prey it has to catch it. This involves coordinating the contraction of the muscle so the predator can run, there must then be an increased blood supply to the muscles to provide them with oxygen and nutrients. At the same time the respiration rate must increase to supply the oxygen and remove the carbon dioxide produced as a result of this increased activity.

Once the prey has been caught and eaten, the digestive system must be activated to digest it. The adjustment of an animal's response to changes in the environment and the complex linking of the various processes in the body that this response involves are called co-ordination. Two systems are involved in co-ordination in animals. These are the nervous and endocrine systems. The first operates via electrical impulses along nerve fibres and the second by releasing special chemicals or hormones into the bloodstream from glands.

FUNCTIONS OF THE NERVOUS SYSTEM

The nervous system has three basic functions:

1. *Sensory function*: To sense changes (known as stimuli) both outside and within the body. For example the

eyes sense changes in light and the ear responds to sound waves. Inside the body, tretch receptors in the stomach indicate when it is full and chemical receptors in the blood vessels monitor the acidity of the blood.

2. *Integrative function*: Processing the information received from the sense organs. The impulses from these organs are analysed and stored as memory. The many different impulses from different sources are sorted, synchronised and co-ordinated and the appropriate response initiated. The power to integrate, remember and apply experience gives higher animals much of their superiority.
3. *Motor function*: The third function is the response to the stimuli that causes muscles to contract or glands to secrete.

All nervous tissue is made up of nerve cells or neurons. These transmit high-speed signals called nerve impulses. Nerve impulses can be thought of as being similar to an electric current.

THE NEURON

Neurons are cells that have been adapted to carry nerve impulses. A typical neuron has a cell body containing a nucleus, one or more branching filaments called dendrites which conduct nerve impulses towards the cell body and one long fibre, an axon, that carries the impulses away from it. Many axons have a sheath of fatty material called myelin surrounding them. This speeds up the rate at which the nerve impulses travel along the nerve. The cell body of neurons is usually located in the brain or spinal cord while the axon extends the whole distance to the organ that it supplies. The neuron carrying impulses from the spinal cord to the hind leg or tail of a horse, for example, can be several feet long. A nerve is a bundle of axons.

A sensory neuron is a nerve cell that transmits impulses from a sense receptor such as those in the eye or ear to the brain or spinal cord. A motor neuron is a nerve cell that transmits impulses from the brain or spinal cord to a muscle or gland. A relay neuron connects sensory and motor neurons and is found in the brain or spinal cord.

CONNECTIONS BETWEEN NEURONS

The connection between adjacent neurons is called a synapse. The two nerve cells do not actually touch here for there is a microscopic space between them. The electrical impulse in the neurone before the synapse stimulates the production of chemicals called neurotransmitters (such as acetylcholine), which are secreted into the gap. The neurotransmitter chemicals diffuse across the gap and when they contact the membrane of the next nerve cell they stimulate a new nervous impulse. After the impulse has passed the chemical is destroyed and the synapse is ready to receive the next nerve impulse.

REFLEXES

A reflex is a rapid automatic response to a stimulus. When you accidentally touch a hot object and automatically jerk your hand away, this is a reflex action. It happens without you having to think about it! Animals automatically blink when an object approaches the eye and cats twist their bodies in the air when falling so they land on their paws. Swallowing, sneezing, and the constriction of the pupil of the eye in bright light are also all reflex actions.

The path taken by the nerve impulses in a reflex is called a reflex arc. Most reflex arcs involve only three neurons. The stimulus (a pin in the paw) stimulates the pain receptors of the skin, which initiate an impulse in a sensory neuron. This travels to the spinal cord where it passes, by means of a synapse, to a connecting neuron called the relay neuron situated in the spinal cord. The relay neuron in turn makes a synapse with one or more motor neurons that transmit the impulse to the muscles of the limb causing them to contract and remove the paw from the sharp object. Reflexes do not require involvement of the brain although you are aware of what is happening and can, in some instances, prevent them happening. Animals are born with their reflexes.

CONDITIONED REFLEXES

In most reflexes the stimulus and response are related. For example the presence of food in the mouth causes the salivary glands to release saliva. However, it is possible to train animals (and humans) to respond to different and often quite irrelevant stimuli. This is called a conditioned reflex.

A Russian biologist called Pavlov carried out the classic experiment to demonstrate such a reflex when he conditioned dogs to salivate at the sound of a bell ringing. Almost every pet owner can identify reflexes they have conditioned in their animals. Perhaps you have trained your cat to associate food with the opening of the fridge door or accustomed your dog to the routines you go through before taking them for a walk.

PARTS OF THE NERVOUS SYSTEM

When we describe the nervous system of vertebrates we usually divide it into two parts:

1. The central nervous system (CNS) which consists of the brain and spinal cord.
2. The peripheral nervous system (PNS) which consists of the nerves that connect to the brain and spinal cord (cranial and spinal nerves) as well as the autonomic (or involuntary) nervous system.

THE CENTRAL NERVOUS SYSTEM

The central nervous system consists of the brain and spinal cord. It acts as a kind of 'telephone exchange' where a vast number of cross connections are made. When you look at the brain or spinal cord some regions appear creamy white (white matter) and others appear grey (grey matter). White matter consists of masses of nerve axons and the grey matter consists of the nerve cells. In the brain the grey matter is on the outside and in the spinal cord it is on the inside.

THE BRAIN

The major part of the brain lies protected within the sturdy "box" of skull called the cranium. Surrounding the fragile brain tissue (and spinal cord) are protective membranes called the meninges, and a crystal-clear fluid called cerebrospinal fluid, which protects and nourishes the brain tissue. This fluid also fills four cavities or ventricles that lie within the brain. Brain tissue is extremely active and, even when an animal is resting, it uses up to 20% of the oxygen taken into the body by the lungs. The carotid artery, a branch off the dorsal aorta, supplies it with the oxygen and nutrients it requires. Brain damage occurs if brain tissue is deprived of oxygen for only 4-8 minutes.

The brain consists of three major regions:

1. The fore brain which includes the cerebral hemispheres, hypothalamus and pituitary gland;
2. The hind brain or brain stem, contains the medulla oblongata and pons and
3. The cerebellum or "little brain".

MAPPING THE BRAIN

In humans and some animals the functions of the different regions of the cerebral cortex have been mapped.

THE FOREBRAIN

The cerebral hemispheres are the masses of brain tissue that sit on the top of the brain. The surface is folded into ridges and furrows called sulci (singular sulcus). They make this part of the brain look rather like a very large walnut kernel. The two hemispheres are separated by a deep groove although they are connected internally by a thick bundle of nerve fibres.

The outer layer of each hemisphere is called the cerebral cortex and this is where the main functions of the cerebral hemispheres are carried out. The cerebral cortex is large and convoluted in mammals compared to other vertebrates and largest of all in humans because this is where the so-called "higher centres" concerned with memory, learning, reasoning and intelligence are situated. Nerves from the eyes, ears, nose and skin bring sensory impulses to the cortex where they are interpreted. Appropriate voluntary movements are initiated here in the light of the memories of past events. Different regions of the cortex are responsible for particular sensory and motor functions, eg. vision, hearing, taste, smell, or moving the fore-limbs, hind-limbs or tail. For example, when a dog sniffs a scent, sensory impulses from the organ of smell in the nose pass via the olfactory (smelling) nerve to the olfactory centres of the cerebral hemispheres where the impulses are interpreted and co-ordinated. In humans and some animals the functions of the different regions of the cerebral cortex have been mapped. The hypothalamus is situated at the base of the brain and is connected by a "stalk" to the pituitary gland, the "master" hormone-producing gland.

The hypothalamus can be thought of as the bridge between the nervous and endocrine (hormone producing) systems. It produces some of the hormones that are released from the pituitary gland and controls the release of others from it. It is also an important centre for controlling the internal environment of the animal and therefore maintaining homeostasis. For example, it helps regulate the movement of food through the gut and the temperature, blood pressure and concentration of the blood. It is also responsible for the feeling of being hungry or thirsty and it controls sleep patterns and sex drive.

THE HINDBRAIN

The medulla oblongata is at the base of the brain and is a continuation of the spinal cord. It carries all signals between the spinal cord and the brain and contains centres that control vital body functions like the basic rhythm of breathing, the rate of the heartbeat and the activities of the gut. The medulla oblongata also co-ordinates swallowing, vomiting, coughing and sneezing.

THE CEREBELLUM

The cerebellum (little brain) looks rather like a smaller version of the cerebral hemispheres attached to the back of the brain. It receives impulses from the organ of balance (vestibular organ) in the inner ear and from stretch receptors in the muscles and tendons. By co-ordinating these it regulates muscle contraction during walking and running and helps maintain the posture and balance of the animal. When the cerebellum malfunctions it causes a tremor and uncoordinated movement.

THE SPINAL CORD

The spinal cord is a cable of nerve tissue that passes down the channel in the vertebrae from the hindbrain to the end of the tail. It becomes progressively smaller as paired spinal nerves pass out of the cord to parts of the body. Protective membranes or meninges cover the cord and these enclose cerebral spinal fluid. If you cut across the spinal cord you can see that it consists of white matter on the outside and grey matter in the shape of an H or butterfly on the inside.

THE PERIPHERAL NERVOUS SYSTEM

The peripheral nervous system consists of nerves that are connected to the brain (cranial nerves), and nerves that are connected to the spinal cord (spinal nerves). The autonomic nervous system is also part of the peripheral nervous system.

CRANIAL NERVES

There are twelve pairs of cranial nerves that come from the brain. Each passes through a hole in the cranium (brain case).

The most important of these are the olfactory, optic, acoustic and vagus nerves:

- *The olfactory nerves*: (Smell) Carry impulses from the olfactory organ of the nose to the brain.
- *The optic nerves*: (Sight) Carry impulses from the retina of the eye to the brain.
- *The auditory (acoustic) nerves*: (Hearing) Carry impulses from the cochlear of the inner ear to the brain.
- *The vagus nerve*: Controls the muscles that bring about swallowing. It also controls the muscles of the heart, airways, lungs, stomach and intestines.

SPINAL NERVES

Spinal nerves connect the spinal cord to sense organs, muscles and glands in the body. Pairs of spinal nerves leave the spinal cord and emerge between each pair of adjacent vertebrae. The sciatic nerve is the largest spinal nerve in the body. It leaves the spinal cord as several nerves that join to form a flat band of nervous tissue. It passes down the thigh towards the hind leg where it gives off branches to the various muscles of this limb.

THE AUTONOMIC NERVOUS SYSTEM

The autonomic nervous system controls internal body functions that are not under conscious control. For example when a prey animal is chased by a predator the autonomic nervous system automatically increases the rate of breathing and the heartbeat. It dilates the blood vessels that carry blood to the muscles, releases glucose from the liver, and makes other adjustments to provide for the sudden increase in activity. When the animal has escaped and is safe once again the nervous system slows down all these processes and resumes all the normal body activities like the digestion of food.

The nerves of the autonomic nervous system originate in the spinal cord and pass out between the vertebrae to serve the various organs. There are two main parts to the autonomic nervous system—the sympathetic system and the parasympathetic system. The sympathetic system stimulates the "flight, fright, fight" response that allows an animal to face up to an attacker or make a rapid departure. It increases the heart and respiratory rates, as well as the amount of blood flowing to the skeletal muscles while blood flow to less critical regions like the gut and skin is reduced. It also causes the pupils of the eyes to dilate.

Note that the effects of the sympathetic system are similar to the effects of the hormone adrenaline. The parasympathetic system does the opposite to the sympathetic system. It maintains the normal functions of the relaxed body. These are sometimes known as the "housekeeping" functions. It promotes effective digestion, stimulates defaecation and urination and maintains a regular heartbeat and rate of breathing.

THE SENSE ORGANS

Sense organs allow animals to sense changes in the environment around them and in their bodies so that they can respond appropriately. They enable animals to avoid hostile environments, sense the presence of predators and find food. Animals can sense a wide range of stimuli that includes, touch, pressure, pain, temperature, chemicals, light, sound, movement and position of the body. Some animals can sense electric and magnetic fields. All sense organs respond to stimuli by producing nerve impulses that travel to the brain via a sensory nerve. The impulses are then processed and interpreted in the brain as pain, sight, sound, taste etc.

The senses are often divided into two groups:

1. The general senses of touch, pressure, pain and temperature that are distributed fairly evenly through the skin. Some are found in muscles and within joints.
2. The special senses which include the senses of smell, taste, sight, hearing and balance. The special sense organs may be quite complex in structure.

TOUCH AND PRESSURE

Within the dermis of the skin are numerous modified nerve endings that are sensitive to touch and pressure. The roots of hairs may also be well supplied with sensory receptors that inform the animal that it is in contact with an object. Whiskers are specially modified hairs.

PAIN

Receptors that sense pain are found in almost every tissue of the body. They tell the animal that tissues are dangerously hot, cold, compressed or stretched or that there is not enough blood flowing in them. The animal may then be able to respond and protect itself from further damage

AWARENESS OF LIMB POSITION

There are sense organs in the muscles, tendons and joints that send continuous impulses to the brain that tell it where each limb is. This information allows the animal to place its limbs accurately and know their exact position without having to watch them.

SMELL

Animals use the sense of smell to locate food, mark territory, identify their own offspring and the presence and sexual condition of a potential mate. The organ of smell (olfactory organ) is located in the nose and responds to chemicals in the air. It consists of modified nerve cells that have several tiny hairs on the surface.

These emerge from the epithelium on the roof of the nose cavity into the mucus that lines it. As the animal breathes, chemicals in the air dissolve in the mucus. When the sense cell responds to a particular molecule, it fires an impulse that travels along the olfactory nerve to the brain where it is interpreted as an odour.

The olfactory sense in humans is rudimentary compared to that of many animals. Carnivores that hunt have a very highly developed sensitivity to scents. For example a polar bear can smell out a dead seal 20 kms away and a bloodhound can distinguish between the trails of different people although it may sometimes be confused by the criss-crossing trail of identical twins. Snakes and lizards detect odours by means of Jacobson's organ. This is situated on the roof of the mouth and consists of pits containing sensory cells. When snakes flick out their forked tongues they are smelling the air by carrying the molecules in it to the Jacobson's organ.

TASTE

The sense of taste allows animals to detect and identify dissolved chemicals. In reptiles, birds, and mammals the taste receptors (taste buds) are found mainly to the upper surface of the tongue. They consist of pits containing sensory cells arranged rather like the segments of an orange. Each receptor cell has a tiny "hair" that projects into the saliva to sense the chemicals dissolved in it.

The sense of taste is quite restricted. Humans can only distinguish four different tastes (sweet, sour, bitter and salt) and what we normally think of as taste is mainly the sense of smell. Food is quite tasteless when the nose is blocked and cats often refuse to eat when this happens.

SIGHT

The eyes are the organs of sight. They consist of spherical eyeballs situated in deep depressions in the skull called the orbits. They are attached to the wall of the orbit by six muscles, which move the eyeball. Upper and lower eyelids cover the eyes during sleep and protect them from foreign objects or too much light, and spread the tears over their surface.

The nictitating membrane or hawis a transparent sheet that moves sideways across the eye from the inner corner, cleansing and moistening the cornea without shutting out the light. It is found in birds, crocodiles, frogs and fish as well as marsupials like the kangaroo. It is rare in mamals but can be seen in cats and dogs by gently opening the eye when it is asleep. Eyelashes also protect the eyes from the sun and foreign objects.

STRUCTURE OF THE EYE

Lining the eyelids and covering the front of the eyeball is a thin epithelium called the conjunctiva. Conjunctivitis is inflammation of this membrane.

Tear glands that open just under the top eyelid secrete a salty solution that keeps the exposed part of the eye moist, washes away dust and contains an enzyme that destroys bacteria. The wall of the eyeball is composed of three layers. From the outside these are the sclera, the choroid and the retina. The sclera is a tough fibrous layer that protects the eyeball and gives it rigidity.

At the front of the eye the sclera is visible as the "white" of the eye, which is modified as the transparent cornea through which the light rays have to pass to enter the eye. The cornea helps focus light that enters the eye. The choroid lies beneath the sclera. It contains a network of blood vessels that supply the eye with oxygen and nutrients. Its inner surface is highly pigmented and absorbs stray light rays. In nocturnal animals like the cat and possum this highly pigmented layer reflects light as a means of conserving light. This is what makes them shine when caught in car headlights. At the front of the eye the choroid becomes the iris. This is the coloured part of the eye that controls the amount of light entering the pupil of the eye. In dim light the pupil is wide open so as much light as possible enters while in bright light the pupils contract to protect the retina from damage by excess light.The pupil in most animals is circular but in many nocturnal animals it is a slit that can close completely. This helps protect the extra-sensitive light sensing tissues of animals like the cat and possum from bright sunlight. The inner layer lining the inside of the eye is the retina. This contains the light sensing cells called rods and cones. The rod cells are long and fat and are sensitive to dim light but cannot detect colour. They contain large amounts of a pigment that changes when exposed to light.

This pigment comes from vitamin A found in carrots etc. A deficiency of this vitamin causes night blindness. The cone cells provide colour vision and allow animals to see details. Most are found in the centre of the retina and they are most densely concentrated in a small area called the fovea. This is the area of sharpest vision, where the words you are reading at this moment are focussed on your retina. The nerve fibres from the cells of the retina join and leave the eye via the optic nerve. There are no rods or cones here and it is a blind spot. The optic nerve passes through the back of the orbit and enters the brain. The lens is situated just behind the pupil and the iris. It is a crystalline structure with no blood vessels and is held in position by a ligament.

This is attached to a muscle, which changes the shape of the lens so both near and distant objects can be focussed by the eye. This ability to change the focus of the lens is called accommodation. In many mammals the muscles that bring about accommodation are poorly developed, Rats, cows and dogs, for example, are thought to be unable to focus clearly on near objects. In old age and certain diseases the lens may become cloudy resulting in blurred vision.

This is called a cataract. Within the eyeball are two cavities, the anterior and posterior chambers, separated by the lens. They contain fluids the aqueous and vitreous humoursr-espectively, that maintain the shape of the eyeball and help press the retina firmly against the choroid so clear images are seen.

HOW THE EYE SEES

Eyes work quite like a camera. Light rays from an object enter the eye and are focused on the retina (the "film") at the back of the eye. The cornea, the lens and the fluid within the eye all help to focus the light. They do this by bending the light rays so that light from the object falls on the retina. This bending of light is called refraction. The light stimulates the light sensitive cells of the retina and nerve impulses are produced that pass down the optic nerve to the brain.

As mentioned before, the retina has two different kinds of cells that are stimulated by light - rods and cones. In humans and higher primates like baboons and gorillas the rods function in dim light and do not perceive colour, while the cones are stimulated by bright light and perceive details and colour. Other mammals have very few cones in their retinas and it is believed that they see no or only a limited range of colour. It is, of course, difficult to find out exactly what animals do see. It is thought that deer, rats, and rabbits and nocturnal animals like the cat are colour-blind, and dogs probably see green and blue. Some fish and most birds seem to have better colour vision than humans and they use colour, often very vivid ones, for recognizing each other as well as for courtship and protection.

HEARING

Animals use the sense of hearing for many different purposes. It is used to sense danger and enemies, to detect prey, to identify prospective mates and to communicate within social groups. Some animals (*e.g.* most bats and dolphins) use sound to "see" by echolocation. By sending out a cry and interpreting the echo, they sense obstacles or potential prey.

BINOCULAR VISION

Animals like cats that hunt have eyes placed on the front of the head in such a way that both eyes see the same wide area but from slightly different angles. This is called binocular vision. Its main advantage is that it enables the animals to estimate the distance to the prey so they can chase it and pounce accurately. In contrast plant-eating prey animals like the rabbit and deer need to have a wide panoramic view so they can see predators approaching. They therefore have eyes placed on the side of the head, each with its own field of vision. They have only a very small area of binocular vision in front of the head but are extremely sensitive to movement.

STRUCTURE OF THE EAR

Most of the ear, the organ of hearing, is hidden from view within the boney skull. It consists of three main regions: the outer ear, the middle ear and the inner ear. The outer ear consists of an ear canal leading inwards to a thin membrane known as the eardrumor tympanic membrane that stretches across the canal. Many animals have an external ear flap or pinna to collect and funnel the sound into the ear canal. The pinnae (plural of pinna) usually face forwards on the head but many animals can swivel them towards the source of the sound.

In dogs the ear canal is long and bent and often traps wax or provides an ideal habitat for mites, yeast and bacteria. The middle ear consists of a cavity in the skull that is connected to the pharynx (throat) by a long narrow tube called the Eustachian tube. This links the middle ear to the outside air so that the air pressure on both sides of the eardrum can be kept the same. Everyone knows the uncomfortable feeling (and affected hearing) that occurs when you drive down a steep hill and the unequal air pressures on the two sides of the eardrum cause it to distort. The discomfort is relieved when you swallow because the Eustachian tubes open and the pressure on either side equalises. Within the cavity of the middle ear are three of the smallest bones in the body, the auditory ossicles. They are known as the hammer, the anvil and the stirrup because of their resemblance to the shape of these objects. These tiny bones articulate (move against) each other and transfer the vibrations of the eardrum to the membrane covering the opening to the inner ear.

The inner ear is a complicated series of fluid-filled tubes imbedded in the bone of the skull. It consists of two main parts.

These are the cochlea where sound waves are converted to nerve impulses and the vestibular organ that is associated with the sense of balance and has no role in hearing. The cochlea looks rather like a coiled up snail shell. Within it there are specialised cells with fine hairs on their surface that respond to the movement of the fluid within the cochlea by producing nervous impulses that travel to the brain along the auditory nerve.

HOW THE EAR HEARS

Sound waves can be thought of as vibrations in the air. They are collected by the ear pinna and pass down the ear canal where they cause the eardrum to vibrate. The vibration of the eardrum sets the three tiny bones in the middle ear moving against each other so that the vibration is transferred to the membrane covering the opening to the inner ear.

As well as transferring the vibration, the tiny ear bones also amplify it. The three tiny bones are called the stirrup, anvil, and hammer. They were called such of their form. In the human ear this amplification is about 20 times while in desert-dwelling animals like the kangaroo rat it is 100 times. This acute hearing warns them of the approach of predators like owls and snakes, even in the dark.

The vibration causes waves in the fluid in the inner ear that pass down the cochlea. These waves stimulate the tiny hair cells to produce nerve impulses that travel via the auditory nerve to the cerebral cortex of the brain where they are interpreted as sound. The hair cells generate a nerve impulse that travels down the auditory nerve to the brain. Remember that sound waves do not pass along the Eustachian tube. Its function is to equalise the air pressure on either side of the tympanic membrane.

BALANCE

The vestibular organ of the inner ear helps an animal maintain its posture and keep balanced by monitoring the movement and position of the head. It consists of two structures - the semicircular canals and the otolith organs. The semicircular canals respond to movement of the body. They tell an animal whether it is moving up or down, left or right. They consist of three canals set in three different planes at right angles to each other so that movement in any direction can be registered. The canals contain fluid and sense cells with fine hairs that project into the fluid. When the head moves the fluid swirls in the canals and stimulates the hair cells. These send nerve impulses along the vestibular nerve to the cerebellum.

Note that the semicircular canals register acceleration and deceleration as well as changes in direction but do not respond to movement that is at a constant speed. The otolith organs are sometimes known as gravity receptors. They tell you if your head is tilted or if you are standing on your head. They consist of bulges at the base of the semi circular canals that contain hair cells that are covered by a mass of jelly containing tiny pieces of chalk called otoliths. When the head is tilted, or moved suddenly, the otoliths pull on the hair cells, which produce a nerve impulse.

This travels down the vestibular nerve to the cerebellum. By coordinating the nerve impulses from the semicircular canals and otolith organs the cerebellum helps the animal keep its balance. The field of animal physiology extends the tools and methods of human physiology to non-human animal species. Plant physiology also borrows techniques from both fields. Its scope of subjects is at least as diverse as the tree of life itself. Due to this diversity of subjects, research in animal physiology tends to concentrate on understanding how physiological traits changed throughout the evolutionary history of animals.

Other major branches of scientific study that have grown out of physiology research include biochemistry, biophysics, paleobiology, biomechanics, and pharmacology. Although the various branches of human knowledge cannot be sharply delimited from each other, it has been found expedient for practical purposes to distinguish one field of study from another. Thus chemistry, although close to physics, is set apart from it, and within the broad field of chemistry one can recognize such special subjects as inorganic chemistry, qualitative analysis, physical chemistry, etc. Not infrequently, as knowledge expands, the older subjects become subdivided. Thus electrochemistry and colloid chemistry are offshoots from physical chemistry. From a broad philosophical point of view some of the distinctions between sciences and their various subdivisions are none too valid. New subjects appear and receive names in treatises or college catalogs, often because some specialist desires to pass on to students information which he knows best or in which he is most interested. In spite of the fact that distinctions between various sciences and their subdivisions may be more or less arbitrary, each branch of knowledge has to a greater or less extent a reason for existence.

Perhaps one can think of each scientific subject as representing an attempt to answer a question. Thus, in biological science, anatomy covers the question of form. It tells us what organisms or their parts look like. Physiology, on the other hand, is concerned with the problem of function. It attempts to describe the actions of various organs or tissues and, more than this, it attempts to explain this action in terms of known chemical and physical forces. In such explanation, moreover, there is an effort to relate the phenomena in terms of cause and effect. Obviously there are various branches of physiological science.

Because of its intimate relation to the science and art of medicine, human physiology has been studied since the very beginning of human learning. Our knowledge of the physiology of other forms of life has been focused on certain types rather than others. Insect physiology is now rather widely studied; plant physiology is an old and well-developed subject. Some groups of organisms have been neglected, and there is comparatively little information about the physiology of certain invertebrate phyla. Sometimes it is convenient to lump together the physiological knowledge of all sorts of organisms into a subject known as "comparative physiology." The point of view of this subject is to emphasize the differences rather than the resemblances between various types of living material. Opinions may differ as to the proper definition of general physiology. To some it may be little more than a diluted vertebrate or human physiology.We should prefer to consider general physiology in the broadest possible sense: it is the science which deals with all forms of living material.

It attempts to answer the question, "What is life?" The general physiologist believes that all life has something in common; his search is often for the least common denominator of living processes. That such a search may not be entirely futile is indicated by the recent history of biological thought. Modern progress in biological science has resulted largely from the fortunate realization that living things, no matter how diverse they appear in externals, are fundamentally much alike. There is an underlying unity which has proved of tremendous importance. Cytologists learned to know cell division and chromosomes from a study of cells best suited for their purposes. Fertilization was studied first in the egg cells of marine invertebrates, and only later was the same process shown to occur in widely different types of organisms.

The fundamental facts of genetics were discovered first for plants, and modern study in this field for the most part contents itself with work on these forms which are easy to raise and which breed rapidly. And yet no one doubts that genetic science is capable of universal application. In physiology, on the other hand, the tendency has been to stress the study of vertebrate or mammalian tissues. Physiology is so close to medicine and its facts are so certain of practical application that there is a strong desire to study animals close to man. The field of mammalian or human physiology has had a marvelous development, and it would be absurd to belittle its striking achievements. And yet, in spite of the fact that students of the physiology of mammals have accumulated a vast store of precise and valuable information, the ultimate mechanisms underlying the activity of mammalian cells and tissues are but little understood. The physiologist has been loath to accept the method of attack so fruitful in other branches of biology.

If students of genetics or cytology had insisted on studying mammals or vertebrates and no other organisms, these sciences would scarcely have developed far. Great advances in theoretical biology are made by study of that type of living material particularly adapted to the problem at hand. Cells of mammals or vertebrates are often more difficult to study than those of simpler forms. The general physiologist, in his search for fundamental information, uses whatever material he can find best suited for his purpose. It is the aim of the general physiologist to discover, in so far as possible, the nature and mechanism of living matter. This is the *raison d'être* of the science.

Its students are more or less convinced that, if one could discover the essential principles which govern the activities and life of any one living cell or living material, the same principles could be used, with minor adjustments and modifications, to explain the life of practically every type of cell. This may seem almost too much to ask, but the progress of biology has repeatedly shown the deep underlying similarity of what seem at first the most diverse manifestations of the living process. The extreme difficulty of biological research is thus to some extent compensated by the underlying unity of living materials.

Although there may be a million different species of animals, there are not a million different biologies, but rather one fundamental science with variations. In general the cell is the unit of living matter, and the life of an organism is to all intents and purposes the sum total of the life of its constituent cells. It seems clear therefore that vital machinery must be interpreted in terms of cells, and general physiology thus becomes cellular physiology. This point of view was strongly emphasized by Verworn, who states over and over again that, if physiology is to attempt fundamental explanation of vital activity, it must be a cell physiology. There are, however, possible exceptions to this point of view. It may be argued that not all living matter is truly cellular. Thus only by a stretch of the imagination can some of the smaller bacteria be regarded as cells; and filtrable viruses which are beyond the limits of microscopic vision, and which may or may not be alive, are certainly not cells in any ordinary sense. This argument seems to have very little force. Biologists generally realise that the cell doctrine is not absolutely true, and that there are forms of living matter which are not truly cellular. Nevertheless, the cell doctrine is true for all practical purposes, and it forms the basis for such sciences as histology, pathology, embryology, etc. Because ultramicroscopic forms of life may exist, neither the histologist nor the embryologist has seen fit to throw over the cell theory. In the same way physicists and physical chemists cling to the second law of thermodynamics, although it is well known that there are special conditions under which the law does not hold. There is also another reason sometimes given for not identifying general physiology with cell physiology.

In investigating the activity of a mass of living matter, it is often much simpler to study great numbers of cells rather than individual units. This is especially true of chemical study, for chemical properties of cells are usually additive. Thus if one is attempting to discover the change in the glycogen content of cells during activity, one can do this best by analyzing great numbers of cells. The analysis is, therefore, conducted on tissues rather than on cells, and in this sense, from the standpoint of method or technic, the study is no longer truly cellular. But the primary interest of the investigation is the glycogen change in the individual cell, so that this argument seems rather a superficial one. At the present time, in many branches of physiology, there is an increasing tendency to study the individual cells in so far as this is possible.

Muscle physiologists are giving more and more attention to the isolated single muscle fibre, and students of nerve are attempting as best they can to isolate single nerve fibres. As a matter of fact, these recent studies of muscle and nerve physiologists have been inspired more by a desire to avoid the complexities which arise from the existence of many fibres not always in perfect time accord, rather than from any desire to study the single fibre from the viewpoint of the student of cell protoplasm. Nevertheless, they indicate a growing realization that the ultimate mechanism responsible for any form of vital activity lies inherent in the individual cells. Whether or not we regard general physiology as synonymous with cell physiology is of no great consequence. A science is whatever its devotees choose to make it. No one could be excluded from the field of general physiology because he did not think in terms of cells.

There is, however, a growing realization that an understanding of cells and cell protoplasm is essential for the interpretation of the various vital processes as they occur in animal and plant organisms. Much of the hope for the future of general physiology lies in the possibility of using simpler types of protoplasm in order to discover the general characteristics of living substance. The field of the general physiologist is as broad as all life itself.

How broad this is, at the present time it is impossible to state, for there is no agreement as to the lower limits of life. Numerous submicroscopic filtrable viruses exist, many of them causing diseases in animals and plants. But whether these viruses are alive or not is a question still being argued. True, they reproduce and some of the larger ones resemble bacteria. On the other hand, unlike bacteria they are incapable of growth on simple synthetic media and they can only exist within bacterial, plant or animal cells.

Moreover, Stanley found that the virus which causes tobacco mosaic disease is a protein molecule, and since then all the plant viruses that have been studied have been found to be protein molecules. However, although some animal viruses seem to consist wholly of protein, others like vaccinia virus contain fat and carbohydrate as well as protein. Viruses vary widely in size and shape. Some are so small that their molecular diameter is only about 25 mμ (0.000,025 mm.).

The largest virus has a volume approximately 10,000 times that of the smallest. Such large viruses can be as large or larger than small bacteria. Some viruses are spherical, some are rod-shaped; one has been found to be tadpole-shaped, with a spherical head and a slender elongate tail. In a study of these diverse, extremely tiny, living or partly living organisms or substances, much can doubtless be learned concerning the nature of life processes. A considerable body of information has already been accumulated. General physiology is a theoretical science just as physical chemistry is. But whereas practical chemists and chemical engineers have long since learned to depend on physicochemical principles, medical scientists and students of applied human physiology and pharmacology have sometimes been slow to realise the relation of their subjects to general physiology.

It is interesting to note that a leading student of pharmacology, which is one of the most practical of biological sciences, has stated that "The manner in which drugs exert their action on cells has become one of the most important fundamental problems of physiology. Fortunately, at the present time, especially among the younger medical men, there is an ever-growing realization of the importance of basic general physiology.

The primary object of the general physiologist is to interpret and explain vital phenomena. Obviously the explanations he seeks must be in terms of known scientific concepts. As best he can, the general physiologist attempts to give a complete description of living substance in terms of its physical and chemical properties; and more than that, he attempts to show exactly what physical and chemical changes occur during vital activity. But mere description is not enough. He must aim to find causal relations between the physical and chemical phenomena observable in the living material and the life itself; in other words he is interested in discovering mechanism. This is his most difficult task.

To measure mechanical, thermal, electrical or chemical change in plant and animal tissues is not too hard, but to find out exactly which changes are primarily responsible for definite phases of vital activity is always exceedingly difficult and often, at our present stage of progress, impossible. Even in the simplest types of vital activity so many changes occur that it is hard to know which ones are intimately responsible for what is happening.

Similar difficulties exist even for non-living machinery. We know how an automobile works because it has been constructed by human hands, but if we were obliged to explain the action of an automobile by what we could see, we might be inclined to believe that its movement was due to the noise it made, the heat it produced or the breeze it generated. As scientists, general physiologists are concerned only with explanations in terms of what is known or what can be known. Whether or not life can be explained in terms of known or knowable chemical and physical forces is a philosophical rather than a scientific question and it should perhaps be left to the philosophers.

For them, and for thoughtful people generally, it is a question of tremendous interest. "Since a man must needs live before he can be a philosopher, no problem of philosophy is more fundamental than the nature of life." For many years, those who believe that eventually explanations of the nature of life processes can be obtained in terms of chemistry and physics have been called mechanists.

On the other hand, those who deny that physical and chemical forces can ever provide a complete explanation have been called vitalists. Between mechanists and vitalists, the arguments have often been bitter—among working biologists generally, the term vitalist is often used as a term of reproach—and hence in recent years there has been a semantic attempt to alter the terminology or to abandon it altogether. But the terms are still properly understood in their original meaning, and it is doubtful if other words or other definitions of the old words will make the arguments any less bitter.

To some modern vitalists, the chief characteristic of living material, the one most difficult to explain chemically or physically, is the extremely complex organization of living systems. The distinguished general physiologist Lillie believes that living material has a "directiveness" quite different from the "randomness" of inanimate material.

The mechanist is impressed with the fact that because of our newer knowledge of viruses there is no longer any sharp and clear-cut distinction between living and non-living, and he also likes to argue that vital phenomena once thought highly mysterious have now been explained, to some extent at least, in physical and chemical terms. To the hard-working physiologist, intent on obtaining at least a partway explanation of some physiological process in physiochemical terms, the validity of the mechanist point of view scarcely seems worth discussing. He is not concerned with the ultimate truth sought by the philosopher. But, unfortunately, when some elderly biologist loses faith in physiochemical interpretation as an answer to all the ultimate problems of biology and leaves his laboratory to publish defeatist or vitalistic views in philosophical journals, these views are promptly read by the philosophers, who by and large have very little acquaintance with biological journals and biological thought and who commonly are convinced that the vast majority of biologists are vitalists.

Nothing could be farther from the truth.In view of the fact that general physiology is a new science, it has never been organized in the same way that older sciences have. There has been no exact standardization of the scope of the subject and still less is there any agreement as to how it can best be approached, or how subdivided.

Anyone may choose his own way of presenting the subject. After a brief survey of the morphology of living cells there will be a discussion of their chemical make-up in so far as it is known, and this will be followed by a consideration of the physical properties of living systems. This preliminary survey of the statics of living systems will precede a more lengthy treatment of the kinetics. In considering the activities of the living machine deal with the chemical aspect of the subject; in other words, the living metabolism.

This will include such topics as nutrition, digestion, respiration, secretion and excretion. The physical aspects of vital activity will then be taken up, and the energy output of the living engine—its production of mechanical energy, heat, electricity and light—will be discussed. The living machine can function only under certain conditions of the environment, and the effect of the environment will next be considered.

Living material can adapt itself to prolonged changes in environment, and we must therefore concern ourselves with the question of acclimatization. An outstanding characteristic of living things is their power to react in special and peculiar ways to sudden changes of the environment, that is to say, to stimulation. This characteristic is known as "irritability," and there is a large body of information concerning various phases of it. After discussing irritability we shall conclude with a brief survey of the phenomenon of aging or senescence, and a short discussion of reproduction.

3

Fish Anatomy

SALINITY

The difference between fatty acid compositions of marine and freshwater fish has been noted by several authors. Some examples of fatty acid patterns Although these fish lipids are higher in w 3 fatty acids, it is clear that freshwater fish have higher levels of w 6 fatty acids than marine species. The average w 6/w 3 ratios are 0.37 and 0.16 for freshwater and marine fish, respectively. Fish in general contain more w 3 than w 6 polyunsaturated fatty acids (PUFA) and should have a higher dietary requirement for w 3 PUFA; thus the dietary EFA requirement of marine fish for w 3 PUFA may be higher than that of freshwater fish.

The same type of difference in the w 6/w 3 ratio between freshwater and seawater is seen when some species of fish migrate from oceans to streams or vice versa. The PUFA ratio of sweet smelt (Plecoglosus altivelis) changes drastically in only one month as they migrate from the sea to a freshwater river. A similar but reverse change occurs in the masu salmon (Oncorhynchus masu) as they migrate from freshwater to seawater. Even within the same species of fish, the salinity of the water seems to cause a dramatic change in the fatty acid pattern. The difference between marine and freshwater fish may be due simply to differences in the fatty acid content in the diet or it may be related to a specific requirement in fish related to physiological adaptations to the environments. The phospholipids are generally considered to be structural or functional lipid, being incorporated to a large extent in the membrane structure of cell and subcellular particles. The triglycerides are more often storage lipids and reflect the fatty acid composition of the diet to a greater extent than do the phospholipids. The fatty acid compositions of the triglyceride and phospholipid fractions of fish lipids are presented. It can be seen that the effect of changing environment on the fatty acid composition of the phospholipid is as great in the case of salmon and considerably greater in the case of the sweet smelt, than it is on the triglyceride composition. Rainbow trout on diets containing either corn oil, which is high in w 6 but low in w 6 PUFA, showed a higher mortality and growth reduction in seawater than in freshwater over the twelve-week feeding period.

Temperature

There are several other factors besides the salinity of the water which affect the fatty acid composition and especially the PUFA of fish. The salmonids are mostly cold-water fish. The fatty acids from a number of marine animals from temperate and arctic waters show some significant differences in the general pattern; unfortunately analysis included fatty acids longer than 20:1. There are a number of other experiments demonstrating the effect of environmental temperature on fatty acid composition of aquatic animals. The general trend towards higher content of long chain PUFA at lower temperatures is quite clear. The ratio decreases with a decrease in temperature.

If the trends in fatty acid composition can be taken as clues to the EFA requirements of fish, the requirement would be greater for fish raised at lower temperatures. Fish raised in warmer waters, such as common carp, channel catfish and tilapia may do better with a mixture of fatty acids.

EFFECTS OF DIET

Some of the fatty acid compositions may be seriously affected by the dietary lipids. The mosquito fish and guppies were fed trout pellets which had an ratio of 2.75. The catfish were fed diets supplemented with either beef tallow or menhaden oil, with ratios of and respectively. These fish were able to alter the dietary ratio in favour of fatty acid incorporation into the flesh lipids even at the highest temperature. Commercially available trout pellets are often low in w 3 PUFA and high in w 6 fatty acids. It is important not to ignore the effect of dietary lipid composition on fatty acid composition of fish fed artificial diets. When the dietary ratio is very high in fatty acids supplied by animal lard or vegetable oils, there is a tendency for fish to alter the ratio of PUFA incorporated in favour of fatty acids. When the dietary oil is a fish oil high in to fatty acids, there is little change in the ratio of lipids incorporated into the fish. This is further suggestive evidence of an EFA requirement of fish for PUFA.

Seasonal Variation

Seasonal variations in the fatty acid composition of fish species have often been reported. Seasonal changes have been observed in total lipid and iodine values of herring oils. The iodine value or degree of unsaturation of the oil was minimal in April and maximal in June. The great increase in unsaturation corresponded to the onset of feeding in spring. The absence of a gas liquid chromatograph (GLC) at the time precluded identification of changes in individual fatty acids.

Flesh and viscera lipid content of the sardine Sardinops melanosticta vary from 3.9 to 10.77 per cent and from 10.9 to 38.3 per cent, respectively. The fatty acids of principal interest with respect to EFA metabolism are 20:4 20:5 and 22:6 3. There was considerable variation in all of these fatty acids in both neutral and polar lipid from both tissues. In the flesh, the 20:4 was consistently higher in the neutral lipid than in the polar lipid. The total consistently higher in polar lipid than in the neutral lipid. Thus, in spite of the major fluctuations in fatty acids caused by changes in diet and temperature throughout the seasons, there was a consistent preferential incorporation of PUFA of the w 3 series into the polar or phospholipid fraction of the lipids. One of the best clues to the EFA requirements of a species can be gained from the fatty acid composition of the lipids incorporated into the offspring or egg. The act of reproduction or spawning also has a significant effect on the seasonal fluctuation of lipids in fish. Fatty acid composition of fish egg lipids is probably distinctive for each species and contains increased levels of compared to the liver lipids of the same female fish. Elevated levels of reduced in the ovary occurred compared to mesenteric fat of Pacific sardine fed a natural copepod diet. The blood fatty acids of the sardine fed the natural diet were similar to those of the ovary.

When the sardines were fed trout food, both the blood and mesenteric fat responded to the diet with elevated 18:2w 6 and reduced 20:5 arid 22:6. The effect of the diet on ovary fatty acid content was considerably less, as relatively high levels of 20:5 and 22:6 were retained. The ovary lipids of the sweet smelt show an increase in 16:0 and a reduction in the PUFA, especially in the phospholipids, compared to the lipids from the flesh of fish caught at the same time of year. The ratio of the ovary was lower than that of the flesh lipids, 0.21 and 0.17 for the ovary compared to 0.31 and 0.20 for the triglycerids and phospholipids of the flesh, respectively.

The hatchability of eggs from common carp fed several different formulated feeds is greatly reduced when the 22:6w3 of the egg lipids is less than 10 per cent. Further, the muscle, plasma and erythrocyte fatty acid compositions are more affected by dietary lipid than those of the eggs. The EFA requirements of a number of species of fish have been investigated in nutritional studies. The fish themselves have given ample evidence for EFA preference by the types of fatty acids they incorporate into their lipids. Fish, in general, tend to utilize ω3 over ω6. This is especially observed when the dietary lipids are high in ω6, as the fish tend to alter the w 6/ω3 ratio towards the ω3 fatty acids in the tissue lipids. The lipids of the egg must satisfy the EFA requirement of the embryo until it is able to feed. The fatty acid composition data suggest that the ω3 requirement is greater in seawater than in freshwater and higher in cold water than in warm water.

BODY LIPID COMPOSITION AND DIETARY LIPID REQUIREMENTS

Detailed information is still lacking on the dietary lipid

requirements of many species of fish, but there is an abundance of information on the fatty acid composition of fish oils. Information on the lipid composition of fish can be used to make some guesses about dietary lipid requirements. Linolenic acid (18:3ω3) resulted in some sparing action and growth promotion in rats and fatty acids of the ω6 EFA prevented all of the EFA-deficiency symptoms. Research with homeothermic land-dwelling animals showed that the ω6 series of fatty acids are the "essential fatty acids", while the ω3 series are considered to be non-essential or only have a partial sparing action on EFA-deficiency.

The series of fatty acids have been shown to be essential to enough species that it began to become accepted that these are the essential fatty acids for all animals. It was assumed by many that fish also required ω6 fatty acids. Many researchers began by supplementing fish diets with vegetable oils, such as corn, peanut, or sunflower oil, which were rich in linoleic acid. The main sympton observed during the development of EFA deficiency in chinook salmon fed fat-free diets was a marked depigmentation that can be prevented by addition of 1 per cent trilinolein, but not by 0.1 per cent linolenic acid. Although the ω6 fatty acids are considered to be essential, one of the general characteristics of fish oils is the low levels of ω6 series fatty acids and the higher levels of ω3 type fatty acids. There is evidence that polyunsaturated fatty acids (PUPA) of the ω3 series, which are present in relatively large concentrations in fish oil, play the role of essential fatty acid for fish.

POLYUNSATURATED FATTY ACID REQUIREMENTS OF FISH

When a test diet containing 13 per cent corn oil and 2 per cent cod-liver oil was fed to rainbow trout, subsequent deletion of the cod-liver oil from the diet depressed growth and produced some kidney degeneration which might be attributed to a lack of sufficient w3 PUFA present in significant quantities in cod-liver oil. Dietary fish oil is superior to corn oil in promoting growth of rainbow trout (Salmo gairdneri) and the yellow-tail (Seriola guingueradiata). Dietary linolenic acid or ethyl linolenate (18:3 ω3) gives a positive growth response for rainbow trout which may be attributed to a dietary requirement for ω3 fatty acids.

NEGATIVE ASPECTS OF LIPIDS IN FISH NUTRITION

The requirement by fish for PUFA of the ω 3 series creates problems with respect to feed storage. These types of fatty acids are very labile on oxidation. The products of lipid oxidation may react with other nutrients such as proteins, vitamins, etc. and reducing the available dietary levels or the oxidation products may be toxic. The effect of oxidized lipids on dietary proteins, enzymes and amino acids have been demonstrated. The use of oxidized menhaden oil in the diets of swine and rats caused decreased appetite, reduced growth, yellowish-brown pigmentation of depot fat and decreased haemoglobin and haematocrit levels.

The negative effects of the oxidized fish oils were reversed by the addition of alpha-tocopherol acetate or ethoxyoquin to the diet. Much of the use of vegetable oils in fish diets in the 1950s and 1960s might, in part, have been based on their greater stability in prepared diets. It has been demonstrated that rancid herring and hake meals in fish feeds caused dark colouration, anaemia, lethargy, brown-yellow pigmented liver, abnormal kidneys and small gill clubbing in chinook salmon. The symptons can be alleviated by addition of alpha-tocopherol to the diets containing rancid fish meals.

The addition of vitamin E would prevent the toxic or negative effects of adding 5 per cent highly oxidized salmon oil to the diet of rainbow trout. This same sparing effect of alpha-tocopherol can also apply to rancid carp feed. The positive nutritional value of ω 3 fatty acids in fish lipids for fish feeds can become a negative factor if adequate care is not taken in the preparation and storage of feeds. Only fresh oils with low peroxide values should be used in feeds. Fish feed ingredients such as fish meals should be protected against oxidation. The level of vitamin E added to the diet should be increased as the PUFA level is increased. The finished feed, if possible, should be stored in air tight containers at reduced temperatures with minimum exposure to UV radiation and other factors accelerating the rate of lipid oxidation. The problems of rancidity or antioxidation of lipids in fish feeds should not be ignored.

Carbohydrates

Carbohydrates represent a broad group of substances which include the sugars, starches, gums and celluloses. The common attributes of carbohydrates are that they contain only the elements carbon, hydrogen and oxygen and that their combustion will yield carbon dioxide plus one or more molecules of Water. The simplest carbohydrates are the three-carbon sugars which figure importantly in intermediary metabolism and the most complex are the naturally occurring polysaccharides, primarily of plant, origin.

In the diet of animals and fish, two classes of polysaccharides are significant:

1. Structural polysaccharides which are digestible by herbivorous species -cellulose, lignin, dextrans, mannans, inulin, pentosans, pectic acids, algic acids, agar and chitin; and
2. Universally digestible polysaccharides-principally starch.

Carbohydrates make up three-fourths of the biomass of plants but are present only in small quantities in the animal body as glycogen, sugars and their derivatives. Glycogen is often referred to as animal starch because it is not present in plants. Derived mono-saccharides such as the sugar acids, amino sugars and the deoxysugars are constituents of all living organisms.

OLIGOSACCHARIDES

The Oligosaccharides raffinose, stachyose and verbascose are present in significant quantities in legume seeds. Raffinose, which is the most widespread among the three, consists of one molecule of glucose linked to a molecule of sucrose at the a - 1,6 position. Its abbreviated chemical name is a -D-Gal. Further chain elongation at the galactose end with another galactose molecule will yield stachyose. These galactose-galactose linkages are all at the a-l,6 position and digestion of these Oligosaccharides by animals requires a highly specific enzyme not elaborated by the animals themselves but by certain bacteria present in the animals guts. The gradual disappearance of oligosaccharides from the cotelydons of legume seeds during germination is part of an intricate process beginning with uptake of water by the seed.

This uptake of moisture releases gibberellic acid which in turn activates the DNA in the seed, thereby triggering the life cycle of the plant. The DNA directs the production of a - galactosidase which is required for the hydrolysis of these Oligosaccharides. Any interference of the DNA transcription process blocks enzyme production and will be evidenced by continued senescence of the seed and persistence of oligosaccharides in the seed cotelydons.

Polysaccharides

The polysaccharides represent a large group of complex carbohydrates which are condensation products of undetermined numbers of sugar molecules. The various subgroups are rather ill-defined and there is a lack of agreement on their classification. Most polysaccharides are insoluble in water. Upon hydrolysis with acids or enzymes they eventually yield their constituent monosaccharides. Starch is a high molecular weight polymer of D-glucose and is the principal reserve carbohydrate in plants. Most starches consist of a mixture of two types of polymers, namely; amylose and amylopectin.

The proportion of amylose and amylopectin is generally one part of amylose and three parts of amylopectin. Enzymes capable of catalyzing the hydrolysis of starch are present in the digestive secretions of animals and fish within their cells. The a-amylases which are found virtually in all living cells cleave the a -D-linkages at random and bring about an eventual total conversion of the starch molecule into the reducing sugars. The principal a -amylases of animal origin are those produced in the salivary gland and the pancreas. Starch is insoluble in water and is stained blue by iodine. Glycogen is the only complex carbohydrate of animal origin. It exists in limited quantities in liver and muscle tissues and acts as a readily available energy source. Dextrins are intermediate compounds resulting from incomplete hydrolysis or digestion of starch.

The presence of a -D-linkages in amylopectin and the inability of a -amylase to cleave these bonds give rise to low molecular weight carbohydrate segments called limit dextrins. These residues are acted upon primarily by acidophilic bacteria in the digestive tract. Cellulose is made up of long chains of glucose units held together by b-D-linkages. The enzymes which cleave these linkages are not ordinarily present in the digestive secretions of animals and fish although some species of shellfish are believed to elaborate cellulase, the enzyme which catalyzes the hydrolysis of cellulose.

Cellulase producing micro-organisms present in the gut of herbivorous animals and fish impart to their host animals the ability to utilize as food the otherwise indigestible cellulose. Other complex polysaccharides in common occurrence are the hemicelluloses and pentosans. Hemicellulose represents a group of carbohydrates including araban, xylan, certain hexosans and polyuronides. These substances are generally less resistant to chemical treatment and undergo some degree of enzymatic hydrolysis during normal digestive processes. Pentosans are polymers of either xylose or arabinose as constituents of plant structural material and vegetable gums, respectively.

CARBOHYDRATE METABOLISM IN FISH

Much of the carbohydrates that enter the diets of animals, including fish, is of plant origin. Carnivorous fish like the Atlantic salmon and the Japanese yellowtail, therefore, deal with little carbohydrate,. Indeed, experiments have shown that these species are ill-equipped to handle significant quantities of raw carbohydrate, in their diets. On the other hand, omnivores such as the common carp and the channel catfish are able to digest fair amounts of carbohydrates in their diets. The grass carp, a herbivore, subsists primarily on a vegetarian diet.

Digestion, Absorption and Storage

The ability of animals to assimilate starch depends on their ability to elaborate amylase. All species of fish have been shown to secrete a -amylase. It has also been demonstrated that activity of this enzyme was greatest in herbivores. In carnivores such as the rainbow trout and sea perch, amylase is primarily of pancreatic origin whereas in herbivores the enzyme is widespread throughout the entire digestive tract. In Tilapia mossambica the pancreas has been shown to be the site of greatest amylase activity followed by the upper intestine. Although the digestion of starch and dextrin by the carnivorous rainbow trout was shown to decrease progressively as levels of the carbohydrates were increased beyond the 20 per cent level, the fish could effectively utilize up to 60 per cent glucose, sucrose or lactose in the diet.

This demonstrates that, contrary to earlier belief, carnivorous fish are capable of efficiently utilizing simple carbohydrate as a primary energy source. The crystalline structure of starch appears also to influence its attack by amylase as evidenced by the two-fold increase in metabolizable energy content of fully cooked (gelatinized) maize in feeding trials with channel catfish. Rainbow trout have also been shown to have a higher tolerance for carbohydrate (present as wheat starch) in the diet when it was cooked.

The process of gelatinization involves both heat and water. If an aqueous suspension of starch is heated, the granules do not change in appearance until a certain critical temperature is reached. At this point some of the starch granules swell and simultaneously lose their crystallinity. The critical temperature is that at which hydrogen bonds of the starch molecule loosen to permit complete hydration, leading to a phenomenon known as "Swelling". Alpha-amylase, promotes a more or less random fragmentation of the starch molecule by hydrolyzing at the a -D- glucosidic bonds in the inner and outer chains of the compound. The result of complete hydrolysis of the amylose component are maltose and D-glucose, while the amylopectin component is reduced to maltose, D-glucose and branched limit dextrins. As a consequence of these action patterns by a -amylase on starch, other enzymes are needed for complete hydrolysis of starch to D-glucose in fish. In this regard, it has been demonstrated that even the carnivorous sea bream possess the ability to digest maltose. On the other hand, cellulase and a -galactosidase have not been shown to be secreted by fish although cellulase of bacterial origin is present in the gut of most species of carps.

The lack of a -galactosidase may partly explain the poor response by fish to dietary soybean meal which contains significant levels of the galactosidic oligosaccharides raffinose and stachyose. As has been pointed out earlier, these oligosaccharides do undergo enzymatic hydrolysis during the germination process to yield galactose and sucrose. It would, therefore, appear that the nutritive value of soybean meal will be enhanced if the bulk of this indigestible starch is first transformed. This can be achieved by soaking the beans for 48 hours prior to processing for meal production.

It should also be pointed out that the nutritive value of pulses and other legume seeds can likewise be improved for fish since oligosaccharides constitute a large portion of the carbohydrates in legume seeds. Data on glucose absorption by fish are scanty. Work with goldfish has shown that active transport of glucose is coupled with Na^+ transport as in most mammals. It is generally believed that absorption takes place on the mucosal surface of intestinal cells. The monosaccharides which result from carbohydrate digestion consist primarily of glucose, fructose, galactose, mannose, xylose and arabinose. Although the rates of absorption of these sugars have been determined for many land mammals, similar information for fish is not available. Glucose does not appear to be a superior energy source for fish over protein or fat although digestible carbohydrates do spare protein for tissue building. Also, unlike in mammals, glycogen is not a significant storage depot of energy despite evidence of an active and reversible Emden-Meyerhoff pathway in fish. The more efficient metabolism of amino acids over glucose for energy could be due to the ability of fish to excrete nitrogenous waste as ammonia from their gills without the high cost of energy in converting the waste to urea.

Other Factors Affecting Metabolism

Apart from genetic adaptation, climatic factors also play an important role in carbohydrate metabolism in fish. Acclimation in fish, in essence, reflects enzyme acclimation, since the animal's ability to survive depends largely upon its ability to carry out normal metabolic functions.

Some enzymes for metabolic acclimation show good compensation while others do not. The enzymes associated with energy liberation (enzymes of glycolysis, pentose shunt, tricarboxylic acid cycle, electron transport and fatty acid oxidation) exhibit temperature compensation whereas, those enzymes dealing largely with the degradation of metabolic products show poor or reverse compensation. It is interesting to note that two key enzymes involved in carbohydrate metabolism, amylase and glucose-6-phosphate dehydrogenase, together with an enzyme involved in fat digestion, lipase, show no temperature compensation. It is not certain if this is in any way connected with the cessation of feeding by fish at low temperatures.

The molecular mechanism of thermal acclimation are not well understood and may consist of changes in synthesis or amounts of a given enzyme. Differences in kinetics, changes in the proportion of isoenzymes suitable for particular temperatures and changes in co-factors such as lipids, co-enzymes, or other factors such as pH and ions may be important in the animal's adjustment to temperature changes.

Energy Transformation

Despite species differences in the tolerance of dietary carbohydrates it is generally believed that the principal end-product of carbohydrate digestion, glucose, is metabolized in a manner prevailing in all cells, *i.e.*, via the reversible Emden-

Meyerhoff pathway. In this pathway, glucose has only one principal fate: phosphorylation to glucose-6-phosphate. Reversible arrows show reaction step or steps catalyzed by same enzymes in both directions.Broken arrows show reactions over many intermediate steps.Paired solid arrows show different enzymes involved in the two directions of the reaction. All transformations proceed with a loss of free energy. Thus, the formation of two moles of lactate from glucose-6-phosphate occurs with free energy change of:

$$DG^{o} = -22000 \text{ cal/mole}$$

The net result is the formation of four molecules of ATP. A functional reversal of this transformation can only occur via a different sequence requiring thc input of six ATP molecules per mole of glucose-6-phosphate recovered.

Cells do not store glucose or glucose-6-phosphate. The readily available storage form is glycogen which is made from glucose-1-phosphate by one pathway and returned by another. Although in mammalian cells glucose-6-phosphate is transformed into fatty acids, such transformation does not appear to take place in fish.

Studies with the common carp indicate that the precursor for lipogenesis is citrate formed when amino acids are actively metabolized through the tricarboxylic acid cycle.

4

Digestive System of Animals

Plant cells are made of organic molecules using energy from the sun. This process is called photosynthesis. Animals rely on these ready-made organic molecules to supply them with their food. Some animals (herbivores) eat plants; some (carnivores) eat the herbivores.

HERBIVORES

Although herbivores eat plant material they have a basic problem. Their own digestive enzymes do not break down the large cellulose molecules in the plant cell walls. Micro-organisms like bacteria, on the other hand, can break them down. Therefore herbivores employ micro-organisms to do the job for them.

There are two types of herbivore:

1. The first, ruminants like cattle, sheep and goats, house these bacteria in a special compartment in the enlarged stomach called the rumen.
2. The second group has an enlarged large intestine and caecum, called a functional caecum, occupied by cellulose digesting micro-organisms. These non-ruminant herbivores include the horse, rabbit and rat.

Even with the help of microorganisms, eating plants still poses problems. Plants are not a good source of nutrients, and herbivores have to eat large quantities of food to obtain all they require. Herbivores like cows, horses and rabbits typically spend much of their day feeding. To give the micro- organisms access to the cellulose molecules, the plant cell walls need to be broken down. This is why herbivores have teeth that are adapted to crush and grind. Their guts also tend to be lengthy and the food takes a long time to pass through it. Eating plants does have some advantages. Plants are immobile so herbivores normally have to spend little energy collecting them. This contrasts with the other main group of animals- the carnivores that often have to chase their prey.

CARNIVORES

Carnivorous animals like those in the cat and dog families, bears, seals, crocodiles and birds of prey catch and eat other animals. They often have to use large amounts of energy finding, stalking, catching and killing their prey. However, they are rewarded by the fact that meat provides a very concentrated source of nutrients. Carnivores in the wild therefore tend to eat distinct meals often with long and irregular intervals between them. Time after feeding is spent digesting and absorbing the food. The guts of carnivores are usually shorter and less complex than those of herbivores because meat is easier to digest than plant material.

Carnivores usually have teeth that are specialised for dealing with flesh, gristle and bone. They have sleek bodies, strong, sharp claws and keen senses of smell, hearing and sight. They are also often cunning, alert and have an aggressive nature.

OMNIVORES

Many animals feed on both animal and vegetable material – they are omnivorous. Most primates including humans belong to this category as do pigs and rats. Their food is diverse, ranging from plant material to animals they have either killed themselves or scavenged from other carnivores. Omnivores lack the specialised teeth and guts of carnivores and herbivores but are often highly intelligent and adaptable reflecting their varied diet.

TREATMENT OF FOOD

Whether an animal eats plants or flesh, the carbohydrates, fats and proteins in the food it eats are generally giant molecules. These need to be split up into smaller ones before they can pass into the blood and enter the cells to be used for energy or to make new cell constituents.

For example:

- Carbohydrates like cellulose, starch, and glycogen need to be split into glucose and other monosaccharides;
- Proteins need to be split into amino acids;
- Fats or lipids need to be split into fatty acids and glycerol.

THE GUT

The digestive tract, alimentary canal or gut is a hollow tube stretching from the mouth to the anus. It is the organ system concerned with the treatment of foods. At the mouth the large food molecules are taken into the gut - this is called ingestion. They must then be broken down into smaller ones by digestive enzymes - digestion, before they can be taken from the gut into the blood stream - absorption. The cells of the body can then use these small molecules - assimilation. The indigestible waste products are eliminated from the body by the act of egestion.

The 4 major functions of the gut are:

1. Transporting the food;
2. Processing the food physically by breaking it up (chewing), mixing, adding fluid etc.
3. Processing the food chemically by adding digestive enzymes to split large food molecules into smaller ones.
4. Absorbing these small molecules into the blood stream so the body can use them.

The regions of a typical mammals gut (for example a cat or dog) are shown in diagram. The food that enters the mouth passes to the oesophagus, then to the stomach, small intestine, caecum, large intestine, rectum and finally undigested material exits at the anus. The liver and pancreas produce secretions that aid digestion and the gall bladder stores bile.

MOUTH

The mouth takes food into the body. The lips hold the food inside the mouth during chewing and allow the baby animal to suck on its mother's teat. In elephants the lips (and nose) have developed into the trunk which is the main food collecting tool. Some mammals, e.g. hamsters, have stretchy cheek pouches that they use to carry food or material to make their nests. The sight or smell of food and its presence in the mouth stimulates the salivary glands to secrete saliva. There are four pairs of these glands in cats and dogs.

The fluid they produce moistens and softens the food making it easier to swallow. It also contains the enzyme, salivary amylase, which starts the digestion of starch. The tongue moves food around the mouth and rolls it into a ball for swallowing. Taste buds are located on the tongue and in dogs and cats it is covered with spiny projections used for grooming and lapping. The cow's tongue is prehensile and wraps around grass to graze it. Swallowing is a complex reflex involving 25 different muscles. It pushes food into the oesophagus and at the same time a small flap of tissue called the epiglottis closes off the windpipe so food doesn't go 'down the wrong way' and choke the animal.

TEETH

Teeth seize, tear and grind food. They are inserted into sockets in the bone and consist of a crown above the gum and root below. The crown is covered with a layer of enamel, the hardest substance in the body. Below this is the dentine, a softer but tough and shock resistant material. At the centre of the tooth is a space filled with pulp which contains blood vessels and nerves. The tooth is cemented into the socket and in most teeth the tip of the root is quite narrow with a small opening for the blood vessels and nerves. In teeth that grow continuously, like the incisors of rodents, the opening remains large and these teeth are called open rooted teeth. Mammals have 2 distinct sets of teeth. The first the milk teeth are replaced by the permanent teeth.

TYPES OF TEETH

All the teeth of fish and reptiles are similar but mammals usually have four different types of teeth. The incisors are the chisel-shaped 'biting off' teeth at the front of the mouth. In rodents and rabbits the incisors never stop growing (open-rooted teeth). They must be worn or ground down continuously by gnawing. They have hard enamel on one surface only so they wear unevenly and maintain their sharp cutting edge. The largest incisors in the animal kingdom are found in elephants, for tusks are actually giant incisors. Sloths have no incisors at all, and sheep have no incisors in the upper jaw. Instead there is a horny pad against which the bottom incisors cut.

The canines or 'wolf-teeth' are long, cone-shaped teeth situated just behind the incisors. They are particularly well developed in the dog and cat families where they are used to hold, stab and kill the prey. The tusks of boars and walruses are large canines while rodents and herbivores like sheep have no (or reduced) canines. In these animals the space where the canines would normally be is called the diastema. In rodents like the rat and beaver it allows the debris from gnawing to be expelled easily. The cheek teeth or premolars and molars crush and grind the food. They are particularly well developed in herbivores where they have complex ridges that form broad grinding surfaces. These are created from alternating bands of hard enamel and softer dentine that wear at different rates. In carnivores the premolars and molars slice against each other like scissors and are called carnassial teeth. They are used for shearing flesh and bone.

DENTAL FORMULA

The numbers of the different kinds of teeth can be expressed in a dental formula. This gives the numbers of incisors, canines, premolars and molars in one half of the mouth. The numbers of these four types of teeth in the left or right half of the upper jaw are written above a horizontal line and the four types of teeth in the right or left half of the lower jaw are written below it.

Thus the dental formula for the sheep is:

0.0.3.3

3.1.3.3

It indicates that in the upper right (or left) half of the jaw there are no incisors or canines (ie there is a diastema), three premolars and three molars. In the lower right (or left) half of the jaw are three incisors, one canine, three premolars and three molars.

The dental formula for a dog is:

3.1.4.2

3.1.4.3

The formula indicates that in the right (or left) half of the upper jaw there are three incisors, one canine, four premolars and two molars. In the right (or left) half of the lower jaw there are three incisors, one canine, four premolars and three molars.

OESOPHAGUS

The oesophagus transports food to the stomach. Food is moved along the oesophagus, as it is along the small and large intestines, by contraction of the smooth muscles in the walls that push the food along rather like toothpaste along a tube. This movement is called peristalsis.

STOMACH

The stomach stores and mixes thefood. Glands in the wall secrete gastric juice that contains enzymes to digest protein and fats as well as hydrochloric acid to make the contents very acidic. The walls of the stomach are very muscular and churn and mix the food with the gastric juice to form a watery mixture called chyme (pronounced kime). Rings of muscle called sphincters at the entrance and exit to the stomach control the movement of food into and out of it.

SMALL INTESTINE

Most of the breakdown of the large food molecules and absorption of the smaller molecules take place in the long and narrow small intestine. The total length varies but it is about 6.5 metres in humans, 21 metres in the horse, 40 metres in the ox and over 150 metres in the blue whale. It is divided into 3 sections: the duodenum (after the stomach), jejunum and ileum.

The duodenum receives 3 different secretions:

1. Bile from the liver;
2. Pancreatic juice from the pancreas and
3. Intestinal juice from glands in the intestinal wall.

These complete the digestion of starch, fats and protein. The products of digestion are absorbed into the blood and lymphatic system through the wall of the intestine, which is lined with tiny finger-like projections called villi that increase the surface area for more efficient absorption.

THE RUMEN

In ruminant herbivores like cows, sheep and antelopes the stomach is highly modified to act as a "fermentation vat". It is divided into four parts. The largest part is called the rumen. In the cow it occupies the entire left half of the abdominal cavity and can hold up to 270 litres. The reticulum is much smaller and has a honeycomb of raised folds on its inner surface. In the camel the reticulum is further modified to store water. The next part is called the omasum with a folded inner surface. Camels have no omasum. The final compartment is called the abomasum. This is the 'true' stomach where muscular walls churn the food and gastric juice is secreted. Ruminants swallow the grass they graze almost without chewing and it passes down the oesophagus to the rumen and reticulum. Here liquid is added and the muscular walls churn the food. These chambers provide the main fermentation vat of the ruminant stomach.

Here bacteria and single-celled animals start to act on the cellulose plant cell walls. These organisms break down the cellulose to smaller molecules that are absorbed to provide the cow or sheep with energy. In the process, the gases methane and carbon dioxide are produced. These cause the "burps" you may hear cows and sheep making. Not only do the micro-organisms break down the cellulose but they also produce the vitamins E, B and K for use by the animal. Their digested bodies provide the ruminant with the majority of its protein requirements. In the wild grazing is a dangerous activity as it exposes the herbivore to predators. They crop the grass as quickly as possible and then when the animal is in a safer place the food in the rumen can be regurgitated to be chewed at the animal's leisure. This is 'chewing the cud' or rumination. The finely ground food may be returned to the rumen for further work by the microorganisms or, if the particles are small enough, it will pass down a special groove in the wall of the oesophagus straight into the omasum. Here the contents are kneaded and water is absorbed before they pass to the abomasum. The abomasum acts as a "proper" stomach and gastric juice is secreted to digest the protein.

LARGE INTESTINE

The large intestine consists of the caecum, colon and rectum. The chyme from the small intestine that enters the colon consists mainly of water and undigested material such as cellulose (fibre or roughage). In omnivores like the pig and humans the main function of the colon is absorption of water to give solid faeces. Bacteria in this part of the gut produce vitamins B and K. The caecum, which forms a dead-end pouch where the small intestine joins the large intestine, is small in pigs and humans and helps water absorption. However, in rabbits, rodents and horses, the caecum is very large and called the functional caecum. It is here that cellulose is digested by micro-organisms. The appendix, a narrow dead end tube at the end of the caecum, is particularly large in primates but seems to have no digestive function.

FUNCTIONAL CAECUM

The caecum in the rabbit, rat and guinea pig is greatly enlarged to provide a "fermentation vat" for micro-organisms to break down the cellulose plant cell walls. This is called a functional caecum. In the horse both the caecum and the colon are enlarged. As in the rumen, the large cellulose molecules are broken down to smaller molecules that can be absorbed. However, the position of the functional caecum after the main areas of digestion and absorption, means it is potentially less effective than the rumen. This means that the small molecules that are produced there can not be absorbed by the gut but pass out in the faeces. The rabbit and rodents (and foals) solve this problem by eating their own faeces so that they pass through the gut a second time and the products of cellulose digestion can be absorbed in the small intestine. Rabbits produce two kinds of faeces. Softer night-time faeces are eaten directly from the anus and the harder pellets you are probably familiar with, that have passed through the gut twice.

THE GUT OF BIRDS

Birds' guts have important differences from mammals' guts. Most obviously, birds have a beak instead of teeth. Beaks are much lighter than teeth and are an adaptation for flight. Imagine a bird trying to take off and fly with a whole set of teeth in its head! At the base of the oesophagus birds have a bag-like structure called a crop. In many birds the crop stores food before it enters the stomach, while in pigeons and doves glands in the crop secretes a special fluid called crop-milk which parent birds regurgitate to feed their young. The stomach is also modified and consists of two compartments.

The first is the true stomach with muscular walls and enzyme secreting glands. The second compartment is the gizzard. In seed eating birds this has very muscular walls and contains pebbles swallowed by the bird to help grind the food. This is the reason why you must always supply a caged bird with grit. In birds of prey like the falcon the walls of the gizzard are much thinner and expand to accommodate large meals.

DIGESTION

During digestion the large food molecules are broken down into smaller molecules by enzymes. The three most important groups of enzymes secreted into the gut are: Amylases that split carbohydrates like starch and glycogen into monosaccharides like glucose.

- Proteases that split proteins into amino acids.
- Lipases that split lipids or fats into fatty acids and glycerol.
- Glands produce various secretions which mix with the food as it passes along the gut.

These secretions include:

Saliva secreted into the mouth from several pairs of salivary glands. Saliva consists mainly of water but contains salts, mucous and salivary amylase. The function of saliva is to lubricate food as it is chewed and swallowed and salivary amylase begins the digestion of starch. Gastric juice secreted into the stomach from glands in its walls. Gastric juice contains pepsin that breaks down protein and hydrochloric acid to produce the acidic conditions under which this enzyme works best. In baby animals rennin to digest milk is also produced in the stomach. Bile produced by the liver. It is stored in the gall bladder and secreted into the duodenum via the bile duct. (Note that the horse, deer, parrot and rat have no gall bladder). Bile is not a digestive enzyme. Its function is to break up large globules of fat into smaller ones so the fat splitting enzymes can gain access the fat molecules.

PANCREATIC JUICE

The pancreas is a gland located near the beginning of the duodenum. In most animals it is large and easily seen but in rodents and rabbits it lies within the membrane linking the loops of the intestine (the mesentery) and is quite difficult to find. Pancreatic juice is produced in the pancreas. It flows into the duodenum and contains amylase for digesting starch, lipase for digesting fats and protease for digesting proteins.

INTESTINAL JUICE

Intestinal juice is produced by glands in the lining of the small intestine. It contains enzymes for digesting disaccharides and proteins as well as mucus and salts to make the contents of the small intestine more alkaline so the enzymes can work.

ABSORPTION

The small molecules produced by digestion are absorbed into the villi of the wall of the small intestine. The tiny finger-like projections of the villi increase the surface area for absorption. Glucose and amino acids pass directly through the wall into the blood stream by diffusion or active transport.

Fatty acids and glycerol enter vessels of the lymphatic system (lacteals) that run up the centre of each villus.

THE LIVER

The liver is situated in the abdominal cavity adjacent to the diaphragm. It is the largest single organ of the body and has over 100 known functions.

Its most important digestive functions are:

- The production of bile to help the digestion of fats (described above) and
- The control of blood sugar levels

Glucose is absorbed into the capillaries of the villi of the intestine. The blood stream takes it directly to the liver via a blood vessel known as the hepatic portal vessel or vein. The liver converts this glucose into glycogen which it stores. When glucose levels are low the liver can convert the glycogen back into glucose. It releases this back into the blood to keep the level of glucose constant. The hormone insulin, produced by special cells in the pancreas, controls this process.

Other functions of the liver include:

- Making vitamin A,
- Making the proteins that are found in the blood plasma (albumin, globulin and fibrinogen),
- Storing iron,
- Removing toxic substances like alcohol and poisons from the blood and converting them to safer substances,
- Producing heat to help maintain the temperature of the body.

In spite of the fact that distinctions between various sciences and their subdivisions may be more or less arbitrary, each branch of knowledge has to a greater or less extent a reason for existence. Perhaps one can think of each scientific subject as representing an attempt to answer a question. Thus, in biological science, anatomy covers the question of form. It tells us what organisms or their parts look like. Physiology, on the other hand, is concerned with the problem of function. It attempts to describe the actions of various organs or tissues and, more than this, it attempts to explain this action in terms of known chemical and physical forces. In such explanation, moreover, there is an effort to relate the phenomena in terms of cause and effect. Obviously there are various branches of physiological science. Because of its intimate relation to the science and art of medicine, human physiology has been studied since the very beginning of human learning. Our knowledge of the physiology of other forms of life has been focused on certain types rather than others. Insect physiology is now rather widely studied; plant physiology is an old and well-developed subject.

Some groups of organisms have been neglected, and there is comparatively little information about the physiology of certain invertebrate phyla. Sometimes it is convenient to lump together the physiological knowledge of all sorts of organisms into a subject known as "comparative physiology." The point of view of this subject is to emphasize the differences rather than the resemblances between various types of living material. Opinions may differ as to the proper definition of general physiology. To some it may be little more than a diluted vertebrate or human physiology. It attempts to answer the question, "What is life?" The general physiologist believes that all life has something in common; his search is often for the least common denominator of living processes.

That such a search may not be entirely futile is indicated by the recent history of biological thought. Modern progress in biological science has resulted largely from the fortunate realization that living things, no matter how diverse they appear in externals, are fundamentally much alike. There is an underlying unity which has proved of tremendous importance. Cytologists learned to know cell division and chromosomes from a study of cells best suited for their purposes. Fertilization was studied first in the egg cells of marine invertebrates, and only later was the same process shown to occur in widely different types of organisms. The fundamental facts of genetics were discovered first for plants, and modern study in this field for the most part contents itself with work on these forms which are easy to raise and which breed rapidly. And yet no one doubts that genetic science is capable of universal application. In physiology, on the other hand, the tendency has been to stress the study of vertebrate or mammalian tissues. Physiology is so close to medicine and its facts are so certain of practical application that there is a strong desire to study animals close to man.

The field of mammalian or human physiology has had a marvelous development, and it would be absurd to belittle its striking achievements. And yet, in spite of the fact that students of the physiology of mammals have accumulated a vast store of precise and valuable information, the ultimate mechanisms underlying the activity of mammalian cells and tissues are but little understood. The physiologist has been loath to accept the method of attack so fruitful in other branches of biology. If students of genetics or cytology had insisted on studying mammals or vertebrates and no other organisms, these sciences would scarcely have developed far. Great advances in theoretical biology are made by study of that type of living material particularly adapted to the problem at hand. Cells of mammals or vertebrates are often more difficult to study than those of simpler forms. The general physiologist, in his search for fundamental information, uses whatever material he can find best suited for his purpose.

It is the aim of the general physiologist to discover, in so far as possible, the nature and mechanism of living matter. This is the *raison d'être* of the science. Its students are more or less convinced that, if one could discover the essential principles which govern the activities and life of any one living cell or living material, the same principles could be used, with minor adjustments and modifications, to explain the life of practically every type of cell. This may seem almost too much to ask, but the progress of biology has repeatedly shown the deep underlying similarity of what seem at first the most diverse manifestations of the living process. The extreme difficulty of biological research is thus to some extent compensated by the underlying unity of living materials. Although there may be a million different species of animals, there are not a million different biologies, but rather one fundamental science with variations. In general the cell is the unit of living matter, and the life of an organism is to all intents and purposes the sum total of the life of its constituent cells. It seems clear therefore that vital machinery must be interpreted in terms of cells, and general physiology thus becomes cellular physiology. This point of view was strongly emphasized by Verworn, who states over and over again that, if physiology is to attempt fundamental explanation of vital activity, it must be a cell physiology. There are, however, possible exceptions to this point of view. It may be argued that not all living matter is truly cellular. Thus only by a stretch of the imagination can some of the smaller bacteria be regarded as cells; and filtrable viruses which are beyond the limits of microscopic vision, and which may or may not be alive, are certainly not cells in any ordinary sense.

This argument seems to have very little force. Biologists generally realise that the cell doctrine is not absolutely true, and that there are forms of living matter which are not truly cellular. Nevertheless, the cell doctrine is true for all practical purposes, and it forms the basis for such sciences as histology, pathology, embryology, etc. Because ultramicroscopic forms of life may exist, neither the histologist nor the embryologist has seen fit to throw over the cell theory. In the same way physicists and physical chemists cling to the second law of thermodynamics, although it is well known that there are special conditions under which the law does not hold. There is also another reason sometimes given for not identifying general physiology with cell physiology. In investigating the activity of a mass of living matter, it is often much simpler to study great numbers of cells rather than individual units. This is especially true of chemical study, for chemical properties of cells are usually additive.

Thus if one is attempting to discover the change in the glycogen content of cells during activity, one can do this best by analyzing great numbers of cells. The analysis is, therefore, conducted on tissues rather than on cells, and in this sense, from the standpoint of method or technic, the study is no longer truly cellular. But the primary interest of the investigation is the glycogen change in the individual cell, so that this argument seems rather a superficial one. At the present time, in many branches of physiology, there is an increasing tendency to study the individual cells in so far as this is possible. Muscle physiologists are giving more and more attention to the isolated single muscle fibre, and students of nerve are attempting as best they can to isolate single nerve fibres. As a matter of fact, these recent studies of muscle and nerve physiologists have been inspired more by a desire to avoid the complexities which arise from the existence of many fibres not always in perfect time accord, rather than from any desire to study the single fibre from the viewpoint of the student of cell protoplasm.

Nevertheless, they indicate a growing realization that the ultimate mechanism responsible for any form of vital activity lies inherent in the individual cells. Whether or not we regard general physiology as synonymous with cell physiology is of no great consequence. A science is whatever its devotees choose to make it. No one could be excluded from the field of general physiology because he did not think in terms of cells. There is, however, a growing realization that an understanding of cells and cell protoplasm is essential for the interpretation of the various vital processes as they occur in animal and plant organisms. Much of the hope for the future of general physiology lies in the possibility of using simpler types of protoplasm in order to discover the general characteristics of living substance. The field of the general physiologist is as broad as all life itself.

How broad this is, at the present time it is impossible to state, for there is no agreement as to the lower limits of life. Numerous submicroscopic filtrable viruses exist, many of them causing diseases in animals and plants. But whether these viruses are alive or not is a question still being argued. True, they reproduce and some of the larger ones resemble bacteria. On the other hand, unlike bacteria they are incapable of growth on simple synthetic media and they can only exist within bacterial, plant or animal cells. Moreover, Stanley found that the virus which causes tobacco mosaic disease is a protein molecule, and since then all the plant viruses that have been studied have been found to be protein molecules. However, although some animal viruses seem to consist wholly of protein, others like vaccinia virus contain fat and carbohydrate as well as protein. Viruses vary widely in size and shape. Some are so small that their molecular diameter is only about 25 mm (0.000,025 mm.). The largest virus has a volume approximately 10,000 times that of the smallest. Such large viruses can be as large or larger than small bacteria. Some viruses are spherical, some are rod-shaped; one has been found to be tadpole-shaped, with a spherical head and a slender elongate tail. In a study of these diverse, extremely tiny, living or partly living organisms or substances, much can doubtless be learned concerning the nature of life processes.

A considerable body of information has already been accumulated. General physiology is a theoretical science just as physical chemistry is. But whereas practical chemists and chemical engineers have long since learned to depend on physicochemical principles, medical scientists and students of applied human physiology and pharmacology have sometimes been slow to realise the relation of their subjects to general physiology. It is interesting to note that a leading student of pharmacology, which is one of the most practical of biological sciences, has stated that "The manner in which drugs exert their action on cells has become one of the most important fundamental problems of physiology. Fortunately, at the present time, especially among the younger medical men, there is an ever-growing realization of the importance of basic general physiology. The primary object of the general physiologist is to interpret and explain vital phenomena. Obviously the explanations he seeks must be in terms of known scientific concepts. As best he can, the general physiologist attempts to give a complete description of living substance in terms of its physical and chemical properties; and more than that, he attempts to show exactly what physical and chemical changes occur during vital activity.

But mere description is not enough. He must aim to find causal relations between the physical and chemical phenomena observable in the living material and the life itself; in other words he is interested in discovering mechanism. This is his most difficult task. To measure mechanical, thermal, electrical or chemical change in plant and animal tissues is not too hard, but to find out exactly which changes are primarily responsible for definite phases of vital activity is always exceedingly difficult and often, at our present stage of progress, impossible. Even in the simplest types of vital activity so many changes occur that it is hard to know which ones are intimately responsible for what is happening. Similar difficulties exist even for non-living machinery. We know how an automobile works because it has been constructed by human hands, but if we were obliged to explain the action of an automobile by what we could see, we might be inclined to believe that its movement was due to the noise it made, the heat it produced or the breeze it generated. As scientists, general physiologists are concerned only with explanations in terms of what is known or what can be known.

Whether or not life can be explained in terms of known or knowable chemical and physical forces is a philosophical rather than a scientific question and it should perhaps be left to the philosophers. For them, and for thoughtful people generally, it is a question of tremendous interest. "Since a man must needs live before he can be a philosopher, no problem of philosophy is more fundamental than the nature of life." For many years, those who believe that eventually explanations of the nature of life processes can be obtained in terms of chemistry and physics have been called mechanists.

On the other hand, those who deny that physical and chemical forces can ever provide a complete explanation have been called vitalists. Between mechanists and vitalists, the arguments have often been bitter—among working biologists generally, the term vitalist is often used as a term of reproach—and hence in recent years there has been a semantic attempt to alter the terminology or to abandon it altogether. But the terms are still properly understood in their original meaning, and it is doubtful if other words or other definitions of the old words will make the arguments any less bitter. To some modern vitalists, the chief characteristic of living material, the one most difficult to explain chemically or physically, is the extremely complex organization of living systems. The distinguished general physiologist Lillie believes that living material has a "directiveness" quite different from the "randomness" of inanimate material. The mechanist is impressed with the fact that because of our newer knowledge of viruses there is no longer any sharp and clear-cut distinction between living and non-living, and he also likes to argue that vital phenomena once thought highly mysterious have now been explained, to some extent at least, in physical and chemical terms.

To the hard-working physiologist, intent on obtaining at least a partway explanation of some physiological process in physiochemical terms, the validity of the mechanist point of view scarcely seems worth discussing. He is not concerned with the ultimate truth sought by the philosopher. But, unfortunately, when some elderly biologist loses faith in physiochemical interpretation as an answer to all the ultimate problems of biology and leaves his laboratory to publish defeatist or vitalistic views in philosophical journals, these views are promptly read by the philosophers, who by and large have very little acquaintance with biological journals and biological thought and who commonly are convinced that the vast majority of biologists are vitalists. Nothing could be farther from the truth.

DIGESTION SYSTEM IN SIMPLE ANIMALS

In simple animals like jellyfish and many worms there are no special organs for this purpose. Dissolved oxygen soaks through their skins from the water in which they live, and dissolved carbon dioxide soaks out. In somewhat more complicated creatures like the earth-worm there is a special fluid, the blood, one of whose functions is to carry oxygen from the skin to the internal organs, and carbon dioxide back.

In the most advanced water-dwellers we find special tufts of thin skin, the gills, into which blood is pumped by a heart or hearts, so as to expose it to water with which it may exchange gases. Gills may be naked, as in the young tadpole, but usually, as they are very delicate, they are protected by lids or pouches, as in fish or lobsters, where they lie behind the head on each side. In air-breathing animals there are two quite different kinds of breathing organ. In insects and spiders, air is carried to every part of the body by tiny branching tubes called tracheae, which open by numerous pores, mainly on the sides of the abdomen. In air-breathing vertebrates and molluscs the blood is exposed in the lungs to air which is continually renewed by the act of breathing. Some animals combine several methods. Thus a frog can use its skin alone for breathing so long as it does not need much oxygen, but it normally employs its lungs as well. The human lungs are elastic organs of a spongy texture, consisting of millions of very small cavities which open into stiff tubes called bronchi.

These in turn open into the windpipe in the front of the neck. Since air cannot enter the space between the lung and the chest-wall, the lungs expand when the chest is expanded. This can happen in two different ways. In the first place the diaphragm, a sheet of muscle separating the chest and belly, and bulging upwards into the former, may contract and force the contents of the belly downwards and outwards. This pulls the bases of the lungs down, and draws air into them, or the muscles which lie between the ribs may contract so as to bring the ribs, which normally slope downwards, into a more horizontal position.The breast-bone is thus pushed away from the vertebral column, while at the same time the diameter of the chest from side to side is increased, and the lungs expand to fill the extra volume.

Muscles acting in the opposite direction force the air out again if necessary; generally the elasticity of the lungs is sufficient. The blood flows to and from the heart, as we have seen, through a closed system of tubes known as arteries, veins, and capillaries. It leaves the heart by the arteries, which are comparatively thick-walled, and seldom near the surface, and returns by the veins, which have thinner walls, and often lie just below the skin. It passes from the arteries to the veins by the capillaries, which are too small to be seen with the naked eye, but are found in almost every tissue. The most easily felt arteries in man are those of the wrist and temple, the most easily seen veins those on the back of the hand and foot. Finally, the heart, which pumps the blood round, is a hollow muscle with four chambers, lying between the lungs, and about as large as the two fists together. The blood from all parts of the body except the lungs enters the heart from behind, near the right-hand top corner, and flows into a chamber called the right auricle. This contracts rhythmically (the average rate in a grown man at rest is about seventy times a minute) and forces its contents into a thicker-walled chamber lying below it, the right ventricle. As soon as this is full it contracts in its turn. The blood is prevented from returning into the auricle by a valve, called the tricuspid. It therefore finds its way out through the pulmonary artery, which leads it to the lungs.

The semilunar valves at the base of the artery prevent it from flowing back into the ventricle when this relaxes at the end of its stroke. In the lungs it has to pass through capillaries in the walls of the air-sacs, so that it is only separated from the air by a very fine membrane, and can easily lose its carbon dioxide and take up oxygen. It returns from the lungs by the pulmonary vein, this time to the left auricle, which contracts at the same time as the right auricle, and fills the left ventricle. The left ventricle, which is assisted by valves not unlike those on the right side of the heart, forces the blood into the largest artery, an elastic tube called the aorta, from which it is distributed all over the body by the other arteries. One way of forcing it round through the narrow capillaries would be to have a cistern at the top of the head from which it flowed down again, but this arrangement would clearly not work when its owner lay down.

In reality, a fairly steady pressure is kept up by the fact that the walls of the aorta and the arteries are always stretched, and continue to squeeze the blood along between the strokes of the ventricles. The heart of a grown man at rest delivers about 8 litres per minute, or just over 100 cubic centimetres per beat. During exercise the heart may deliver three or four times as much per minute, mostly as the result of an increased number of beats per minute. The beat of the heart which is felt below the fifth rib on the left side is due to the left ventricle striking the wall of the chest each time it contracts and stiffens. The blood is squirted along the arteries at a rate which may be as high as 50 centimetres a second, and each fresh wave causes a pulse in the artery. If an artery is cut, the blood comes in a series of spurts from the side nearest the heart, and can be stopped by pressure on the heart side of the cut.

It is easy (and safe) to stop the pulse in the wrist by pressing on the same artery higher up, either inside the elbow or inside the upper arm just below the armpit, where it can be felt. The blood flows through the capillaries very slowly, at about half a millimetre per second. Their very thin walls allow water, gases, and dissolved substances to be exchanged between the blood and the tissues with great ease. From the capillaries the blood oozes gently into the veins. They have no pulse, and a comparatively small pressure and rate of flow.

A cut vein bleeds steadily and the flow can be stopped by compressing the side away from the heart. The flow of blood in the veins is assisted by the presence of valves in them, which only allow it to move towards the heart. If the finger-tip be run along one of the veins of the forearm away from the heart, the vein will dilate on the heart side of each valve. When therefore the contraction of various muscles squeezes the veins, the blood can only flow towards the heart. If a man stands quite still the blood tends to accumulate in the veins of his legs, and he is liable to faint from failure of the supply to his brain. When he starts walking the blood is at once squeezed out of these veins.

The only veins in man which do not lead straight back to the heart are those from the digestive canal and some other abdominal organs, which pass into the liver. Here the food absorbed by the blood from the gut is dealt with. As the liver needs oxygen, it has also a supply of fresh arterial blood. In most other animals the circulation is somewhat different. Thus in a fish the blood only goes once through the heart in a complete journey, instead of twice, and it all passes through the gills before going on to the tissues. In the frog, in keeping with its two ways of breathing, we have a condition almost half-way between that found in fish and men. We must now consider how the blood acts as a carrier of oxygen and carbon dioxide. Water at body-temperature will only take up one two-hundredth of its volume of oxygen from air; if the blood were no better than this we should need a heart working forty times as fast as the one we have. Actually the oxygen is carried round in loose chemical combination with a body called haemoglobin. This is a protein, but contains iron as well as the usual carbon, hydrogen, oxygen, nitrogen, and sulphur.

It is of a purple colour, but becomes red on combining with oxygen, which it readily takes up from the air. Thus the blood in the veins is purplish, but if exposed to air either in the lungs or by opening a vein, it at once becomes red. The haemoglobin which it contains enables blood to hold 18 volumes of oxygen per cent., which is 6/7; of the amount contained in the same volume of air, and about 40 times what the blood could carry without haemoglobin. Since blood carries dissolved solids as well as gases, a single set of capillaries supplies the tissues with all that they need for activity, growth, and repair, besides removing most of the waste products. The carbon dioxide produced in the tissues is mainly carried, not in solution as such (CO_2), but in combination as sodium bicarbonate ($NaHCO_3$).

If we look at a drop of blood under a microscope we see that it consists of a clear fluid full of little reddish-yellow biscuit-shaped bodies about 0.007 millimetre in diameter. They can just be seen with a powerful hand-lens. A cubic millimetre of blood contains about five million, and as a man has about 4 litres of blood the total number in his body is about 20 million million, or far more than the number of men who have lived since history began, or probably at all. Their total surface is about 1,500 times the surface of the body. All the haemoglobin of the blood is contained in them, and they therefore carry round the oxygen from the tissues. Their huge surface area renders this exchange easy. They are cells which have lost their nucleus and are therefore not fully alive, but act as passive carriers of oxygen. They are always being produced in large numbers in the bone-marrow, and destroyed in the spleen and liver when worn out. For every 500 or so of these red corpuscles there is one white corpuscle. They are true cells with nuclei, and some of them are capable of active movements. There are at least six different kinds with different functions, mostly of a protective character.

Thus one kind eat up parasites found in the blood, another kind burrow through the walls of capillaries in inflamed areas, and remove dead or injured tissue and disease-germs, often forming collections of 'matter' or pus. Others produce substances which kill disease-germs, and so on. The blood also contains non-cellular bodies called platelets, which are smaller than the corpuscles and are concerned in clotting and in immunity; there may also be tiny drops of oil after a fatty meal. Besides the blood-vessels, all vertebrates possess another system of vessels which open into the veins, but contain clear fluid called lymph.

The lymphatic vessels slowly drain away fluid from the spaces between the cells. On the course of the vessels the fluid passes through small lymph-nodes or glands, which may be felt under the skin in the neck, armpit, or groin, before entering the blood. The lymph contains white corpuscles which are largely produced in these glands, and abnormal bodies from the tissues are dealt with in them. Thus if the arm is inflamed, the lymph-nodes of the armpit are generally enlarged, as they are busy destroying poisons or bacteria from the inflamed tissue. The lymphatics also play a part in digestion, which we shall study later. We must now see how the cells obtain the food-stuffs which they require. In simple animals such as polyps every cell is so close to the digestive cavity that it can obtain food thence directly. In most higher animals, however, the cells which line the alimentary canal and its glands are highly specialized for the purpose of breaking up the food-stuffs into soluble forms and passing them rapidly into the blood or other body-fluids.

This process is called digestion. A few internal parasites like the tapeworm have no gut, but absorb their food through their skin, relying on their host to break it up for them. The course of digestion in man is as follows. The food is chewed in the mouth, where it is also moistened by saliva, a fluid secreted mainly by three neighbouring pairs of glands. Two pairs of these lie between the tongue and the lower jaw, the third pair (whose inflammation causes mumps) lie below the external ear, mainly inside the lower jaw-bone. Besides moistening the food, saliva contains an enzyme, ptyalin, which breaks up starch into an easily soluble sugar called maltose. Enzymes play a much larger part in digestion than do mechanical processes. Each digestive enzyme is a definite substance with the property of bringing about, or enormously speeding up, a particular chemical reaction.

Pepsin from the stomach will split up half a million times its weight of protein, but will not alter starch or fat. So delicately is an enzyme adjusted to the substrate on which it acts, that as most food molecules are asymmetrical (as we know from their rotation of the plane of polarized light and their asymmetrical crystals) so are the enzymes that act on them. Enzymes have been compared to keys which open certain locks only. But they are like 'Yale' rather than ordinary keys. For we can make in the laboratory a sugar or peptide (part of a protein molecule) which only differs from the natural variety in that its molecules are related to the natural molecules as a left hand to a right or an object to its image in a mirror.

Enzymes will not act on these artificial substances, though they digest their mirror-images which occur in nature. So Alice, who went through the looking-glass in the story, could not have digested the looking-glass proteins and carbohydrates. She would have had to get her energy from fat and alcohol, whose molecules are symmetrical, and would finally have died of protein starvation. Enzymes are not alive, and can still work when removed from the body; but they are generally destroyed by boiling. The action of ptyalin can easily be shown. A solution of boiled starch gives a blue colour with iodine. If saliva is allowed to act on it for a few minutes at body-temperature or a few hours in the cold, the starch is broken down to sugar and loses this property. If the saliva is first boiled no change occurs. When the food is swallowed, it is passed into the gullet or oesophagus. To get there it must pass over the mouth of the wind-pipe, and when any falls down this we choke. To prevent this, the breathing is stopped while we swallow, and the cartilages at the top of the wind-pipe which are concerned in voice-production are brought together so as to close the top of the larynx. It is further protected by a cartilaginous lid at the back of the tongue called the epiglottis, which is pressed backwards over it by the food during swallowing. The movements of the laryngeal cartilages can easily be felt. As the food leaves the mouth it passes out of the control of consciousness and will.

The movements of our digestive canal, except at the two ends, are carried out by smooth or involuntary muscle and are controlled by a special part of the nervous system over which the mind does not preside. The food or drink on leaving the mouth is seized by the smooth muscle of the oesophagus, which contracts behind it and relaxes in front, thus passing it rapidly down into the stomach. Owing to this gripping action one can eat or drink while standing on one's head. The narrow duct opens above into the cavity of the stomach, among columnar cells. The main part of the gland is a simple tube, formed of a single layer of cells; these secrete pepsin. A few darker-stained cells are seen near the outer side of the tube; these probably secrete hydrochloric acid. The human stomach is a bag of smooth muscle lined with a membrane consisting mainly of microscopic glands.

When expanded it will generally hold about 2 litres. The gastric juice is a clear fluid containing about ½ per cent. free HCl, and several enzymes, of which the most important is pepsin. In presence of acid (though not in a neutral or alkaline fluid), pepsin causes proteins to break up into bodies called peptones and proteoses, which have very much smaller molecules than the proteins from which they are derived, and are more soluble. Some carbohydrates are attacked by the hydrochloric acid. Thus each molecule of cane-sugar is split into one of glucose and one of fructose, and inulin, a starch-like body found in many plants, is broken up into fructose. Fats are less affected in the stomach. A meal remains in the stomach for a time—generally between one and four hours—which depends on the nature and quantity of the food taken. All this time the muscular walls of the organ are contracting in such a way as to mix the food thoroughly with the gastric juice. Hardly any absorption occurs in the stomach. When its contents are fully mixed the muscular ring surrounding the lower orifice of the stomach, the pylorus, opens, and a jet of its contents is squirted into the duodenum, the top portion of the small intestine. After a heavy meal the stomach may take several hours in emptying itself.

The human small intestine is a tube about 6 metres long and 3 cm. in diameter when relaxed. It is lined by fine projections of the mucous membrane, the villi, which give it a velvety texture, and increase tenfold the surface available for absorption. Between the bases of the villi are the mouths of numerous microscopic glands, and just beyond the pylorus open the ducts of two large glands, the liver and pancreas. The liver, among its other functions, secretes bile. This contains substances which have the property, possessed in a less degree by soap, of lowering the surface-tension of water in which they are dissolved. Drops of oil or melted fat tend to join up so as to have as small a surface as possible. This tendency of the surface to shrink is prevented by the bile in the intestine, so when fat-drops are broken up there they do not coalesce again, but form a milky emulsion. Hence the fat-splitting enzyme made in the pancreas can act on a vastly greater surface of fat than would otherwise be available. A man with jaundice from a blocked bile-duct can digest milk, whose fat is already broken up, but not suet or butter, which form big drops.

The bile also contains pigments formed from the haemoglobin of worn-out red corpuscles. These are excreted in the faeces and give them their yellow colour. In jaundice the bile-duct is blocked, and accordingly the skin becomes yellow and the faeces white. Bile is stored in the gall-bladder till required. The pancreas secretes a juice containing a number of enzymes. One of these, like ptyalin, breaks down starch into maltose. Others break down maltose to glucose, and lactose (milk-sugar) into glucose and galactose, a very similar substance. Another breaks down fat into glycerol and fatty acids, which partly combine with alkali to form soap. In every case the molecules formed will pass more easily through a membrane than those of the food. If the juice contained a protein-splitting enzyme it would probably digest the pancreas itself and its duct. However, it contains a substance which, on mixing with the secretions of the intestinal glands, yields trypsin, an enzyme which attacks proteins and the products of peptic digestion, breaking them down into amino-acids. These enzymes will not act in an acid fluid, so the bile and pancreatic and intestinal juices contain enough sodium bicarbonate to neutralize the acid of the gastric juice.

The secretion of the intestinal glands, besides the substance which helps to form trypsin, contains an enzyme which will break down peptones, though not whole protein molecules, into amino-acids. There are also enzymes which break down milk-sugar and cane-sugar to sugars containing only six carbon atoms. So far as we know, no animal more complicated than a snail produces an enzyme which will break up cellulose. But grass consists very largely of cellulose, and to digest it hoofed animals and other plant-eaters employ bacteria which grow in their digestive canal.

In cud-chewers like the cow and sheep these bacteria live in special compartments of the stomach; in the horse in the large intestine, which may have a capacity of 200 litres. The bacteria can get very little oxygen, so they cannot oxidize the cellulose, but they turn a good deal of it into methane, which is wasted, and leave much undigested. The rest, however, is broken up into small molecules which the animal can absorb.

In man, cellulose is not digested, but it is useful in giving buik to the faeces and preventing constipation, which easily occurs when the food leaves no indigestible residue. The food is thus broken up into easily soluble constituents, and is ready to be absorbed. This is done by the epithelium of the small intestine. The passage through is not a mere filtration. For example, the blood contains one part of sugar per thousand, and if sugar merely filtered across from gut to blood, it could never quite disappear from the gut as it in fact does. Actually, during the absorption of food, the absorptive cells perform work, for which they need an extra supply of oxygen. The fats are, in part at least, put together again from glycerine and soap, and passed as a fine milky emulsion into the lacteals, as the lymphatics of the small gut are called.

These join together to form the thoracic duct, which runs up through the chest and empties into the jugular vein at the base of the neck. The blood may be noticeably milky after a heavy meal of fat. The rest of the products of digestion are passed into the capillaries, and dissolved in the blood. All the veins from the gut run to the liver, and here the food is further dealt with. Thus sugar, if not needed immediately for oxidation, is stored as a starchy body called glycogen, discovered by Claude Bernard, and gradually liberated as required later on. Some of the sugar and most of the fat are stored elsewhere.

Part of the sugar is stored as glycogen in muscles, but much of it is made into fat and stored under the skin and round some of the internal organs, along with fat from the food. In the frog fat is stored in special fat-bodies. The stored glycogen is later split up into glucose for use in the body by an enzyme, which continues to work after death, and the sugar thus liberated gives liver its well-known sweet taste. Absolutely fresh liver is not sweet. Again, ammonia, which is formed in the digestion of many proteins, is, in the liver, mostly combined with carbon dioxide to form urea:

$$2NH_3 + CO_2 = CON_2H_4 + H_2O.$$

Urea is a very innocuous body, but ammonia rather poisonous if it gets to the brain, so that if the blood from the gut is short-circuited into the vena cava instead of going into the liver, a heavy meal of meat may cause convulsions. The liver also deals on the same sort of lines with any excess of amino-acids in the blood and with various poisons; its most important function is thus the regulation of the blood's composition, and not the secretion of bile.

During starvation the body lives on its stores of fat, and the liver takes on the new duty of converting the stable fats such as those of suet into oils like that of linseed, which are very easily oxidized. The unabsorbed residue of the food from the small intestine passes into the large intestine, where it remains in man for a day or so, and is acted on by bacteria. These are of little or no value to man, though very valuable to some animals. In man little but water is absorbed there. By this removal of water the bulk of the waste food is reduced by about nine-tenths. The large intestine also excretes poorly soluble salts, such as calcium phosphate, from the blood.

These would clog up the urinary passages if they were excreted by the kidneys. The active tissues take oxygen, sugar, fat, and aminoacids from the blood, and use them for oxidation, growth, and repair. Into the blood they empty waste-products, of which the most important are water, carbon dioxide, and urea, but other soluble waste-products of protein metabolism include sulphuric and phosphoric acids, creatinine, and uric acid. The last two contain C, H, O, and N.

They are all excreted by the kidney except the carbon dioxide and some of the water, which go out by the lungs, and, in the case of water, the skin. The kidneys consist of a mass of tubules (about a million on each side in man) each beginning in a capsule containing a tuft of capillaries, and ending, after a winding course, in the central cavity of the kidney. The capillary tuft seems to act as a filter, and the fluid that soaks through it is blood minus corpuscles and minus a few of the large molecules, such as the proteins concerned in clotting.

As this filtrate runs down the tubules, the cells lining them reabsorb valuable constituents of the blood, such as sugar, and probably add unwanted ones such as urea, ammonia, uric acid, creatinine, and sulphates. The urine trickles down the ureters into the bladder, whence it is emptied from time to time. An adult man produces on an average about 1.5 litres of urine per day, containing 30 grams of urea, 15 of sodium chloride, and 10 of other soluble waste-products. The blood is thus the medium of exchange between the different parts of the body.

5

Animal Breeding

The science of Animal breeding is defined as the application of the principles of Genetics and biometry to improve the efficiency of production in farm animals. These principles were applied to change animal populations thousands of years before the sciences of genetics and biometry were formally established. The practice of animal breeding dates back to the Neolithic period, when people attempted to domesticate wild species such as reindeer, goats, hogs and dogs. Domestication was performed through controlled mating and reproduction of captive animals which were selected and mated based on their behaviour and temperament. Judging from cave paintings that have survived, selection was also applied to some qualitative traits such as coat colour and the absence or presence of horns.

Without written records, there is no certain knowledge of the evolution of animal breeding practices, but written documents dating back more than 4000 years indicate that humans appreciated the significance of family resemblance in mating systems, recognized the dangers of intense inbreeding, and used castration to prevent the reproduction of undesirable males. Progress in the performance of domesticated animals through these selection practices was very slow; improvements were mainly due to animals adapting better to their environments. Robert Bakewell, an English animal breeder of the 18th century, is considered the founder of systematized animal breeding. He was the first to emphasize the importance of accurate breeding records, introduced the concept of progeny testing to evaluate the genetic potentials of young sires, and applied inbreeding to stabilize desired qualitative traits. He also promoted concepts such as "like begets like," "prepotency is associated with inbreeding" and "breed the best to the best." Bakewell and his contemporaries in Europe pioneered the development of diverse breeds of Beef cattle, Dairy cattle, Sheep. Colour, conformation, geographical origin and some production characteristics were the main factors that differentiated these breeds. Wide geographical redistribution of animal populations was also an important factor in the formation of new breeds, as invading armies, migrating people and traders transported livestock to new lands.

Animal breeding as a modern science belongs to the 20th century. Although numerous geneticists and biometricians have made significant contributions to the development of this science, J.L. Lush of Iowa State University is considered the father of the modern science of animal breeding. Lush and his students developed major scientific procedures applicable to the genetic improvement of farm animals.

SELECTIVE BREEDING

Selective breeding in domesticated animals is the process of a breeder developing a cultivated breed over time, and selecting qualities within individuals of the breed that will be best to pass on to the next generation. Breeding techniques such as inbreeding, linebreeding and outcrossing are utilized by breeders in the maintenance and improvement of their chosen breeds.

Breeding Stock

"Breeding stock" is a term used to describe a group of animals used for purpose of planned breeding. When an individual is looking to breed animals, he or she is looking for certain valuable traits in purebred stock for a certain purpose, or may intend to use some type of crossbreeding to produce a new type of stock with different, and presumably superior abilities in a given area of endeavor.

To breed chickens, a typical breeder intends to receive eggs, meat, and new, young birds for further reproduction.

Thus the breeder has to study different breeds and types of chickens and analyse what can be expected from a certain set of characteristics before he or she starts breeding them. Accordingly, when purchasing initial breeding stock, the breeder seeks a group of birds that will most closely fit the purpose intended.

Purebred Breeding

The idea of *breed purity* may strike an unpleasant chord with some individuals because it is reminiscent of nineteenth-century eugenics notions of a "superior strain" in humans, supposedly exemplified by members of an aristocratic social class or of certain races. The application of theories of eugenics had far-reaching consequences for human beings, particularly in cases such as The Holocaust. However, the idea of a superior strain in animals, such as that incorporated in the breeding of Thoroughbred race horses, was less controversial. By "breeding the best to the best," employing a certain degree of careful inbreeding, considerable culling, and selection for "superior" qualities, one could develop a bloodline or "breed" superior in certain respects to the original base stock or landrace which had been produced primarily by natural selection.

This process forms the basis for a "true-breeding" line of animals, which today is generally called a "purebred" line. Once created, a purified line is "closed" and preserved from dilution and possible "debasement" by unrelated stock. However, this theory is not without some negative consequences. When the ideal of purified lineage or purely aesthetic type is seen as an end in itself, without consideration of usefulness or vigour, a breed can suffer over time. Also, if a breed registry is closed with too small a gene pool, over time, degradation due to inbreeding depression may occur. For many purebred breeders, the need for a periodic outcross is recognized and incorporated into an overall plan. The observable phenomenon of hybrid vigour stands in contrast to the notion of breed purity. However, on the other hand, indiscriminate breeding of crossbred or hybrid animals may also result in degradation of quality, particularly when animals are mass-produced without concern for overall health or longevity.

Backyard Breeding

The term backyard breeder is a general term, sometimes considered derisive, used to describe people who are "home breeders", and breed only a few animals for personal use or occasional sale. However, while some backyard breeders pay strict attention to quality, the term usually describes those who allow animals, particularly dogs or horses, to procreate regardless of physical, genetic, and/or emotional health. It is also used to describe unreputable commercial breeders. In the process of careless breeding, many backyard breeders produce genetically weak animals that can be predisposed to genetic disorders or debilitating physical deformities, particularly those that result from inbreeding.

In dog breeding, these individuals focus on only a limited aspect of the dog, while ignoring the original function and temperament of the breed. When such breeding is carried out on a large scale, the venue is called a *puppy mill* or *puppy farm.* Backyard breeding is popularly blamed for the proliferation of aggressive dogs for the sports of dog baiting and dog fighting. Dog fanciers generally believe that such ill-bred dogs are the reason for the bad public reputation of certain dog breeds, and the increasingly frequent enactment of breed-specific legislation. Because of the time and expense of feeding and caring for horses, which produce one foal per year, there are fewer horse breeders who produce animals *en masse,* though some individuals do engage in animal hoarding and breed far more animals than they can support.

In the horse world, over breeding of grade animals that cannot be sold except for very low prices raises concerns that such animals will be slaughtered for horsemeat. Some, though not all, colour breeders make errors similar to puppy mill breeders by focusing on only one characteristic to the detriment of health, soundness, and quality conformation. In addition, industrial breeders, such as those who produce *pregnant mare urine* for the manufacture of premarin, allow many mares to be bred indiscriminately, simply so that they are able to produce the urine laced with reproductive hormones that forms the base for the drug. Their foals are generally considered an unwanted byproduct, particularly males (colts), and sold off, frequently to the slaughter market.

IN-BREEDING AND OTHER BREEDING METHODS

One of the most hotly talked about topics with regard to pure-bred dogs is the use of in-breeding. This is a term that is often misused and is extremely misunderstood. Part of the misunderstandings comes from differences in the way the terms are used within the scientific/medical field, and how it is commonly used by breeders. These are the most commonly accepted definitions used by serious dog breeders and will be the definitions used within this article.

- *In-breeding:* This is the breeding of closely related animals. Brother-Sister, Parent-Offspring.
- *Line-breeding:* This is the breeding of animals that share common ancestors but are not closely related. The dogs may share a common great-grandparent.
- *Out-cross:* This is generally considered the breeding of animals with no common ancestors within the first 4 or 5 generations.

Common Misconceptions

- *In-breeding causes genetic diseases:* Breeding closely related animals increases the possibility that any bad genes in a line will show up. It does not 'cause' genetic disease.
- *Out-crossed dogs are healthier:* This is only partly true. There is a known phenomenon called Hybrid Vigour. Two animals of unrelated strains breed and the offspring is often bigger and grows faster than it's purebred cousins. This method is often used by farmers in order to get their animals to market sooner. But one of the biggest misconceptions of hybrid vigour is that it applies to all animals of mixed

heritage. Hybrid Vigour only applies to the animals that are the direct offspring of the crossing of the unrelated strains. In other words if you continue to breed animals of different strains there generally will not be any additional increase in hybrid vigour. If the unrelated strains share common genes for genetic disorders, hybrid vigour will not over ride the risk of the disorder showing up. Out-crossing can also cause problems if widely divergent physical types are mixed due to differences in growth rates and bone and muscle sizes.

BENEFITS OF EACH TYPE OF BREEDING

By definition, purebred dogs have a smaller gene pool to draw on than mixed breed dogs. That smaller gene pool gives the breed its individual characteristics, such as physical appearance and temperament. It is what makes a poodle a Poodle and a Golden Retriever a Golden Retriever. But there is considerable controversy with regard to whether the gene pools of the modern pure-bred have become too small.

Inbreeding

In-breeding is more likely to help "set" or "fix" a particular trait within a breed or a line by narrowing the gene pool to favour those traits. So if a breeder is looking to set a particular desirable feature of their line then in-breeding and choosing the offspring most strongly possessing that trait can be beneficial. In the first pedigree Sydney is the maternal son of Annie, thus doubling on Annie's genes. Or another way to look at it is Annie is both Sydney's mother and Grand-mother. In the second example we have a full brother/sister in-breeding so we are doubling up on the genes from both Kiley and Annie.

In-breeding can also help identify those bad genes that exist within a line. Dogs possessing the bad genes can be eliminated from a breeding programme and carriers also identified. Intermittent in-breeding within a line or breed is not damaging to the long term health of the animals. However, in-breeding over successive generations can lead to reduced fitness and fertility problems among the offspring, resulting in phenomena known as In-breeding Depression. It can take many generations to show up depending on the traits involved. To use this method responsibly a breeder would not want to in-breed on animals with known genetic disorders, temperaments not in keeping with its given breed, or known serious structural faults, or to in-breed frequently even on healthy-superior specimens.

Line-Breeding

Line-breeding is another way to help "set" or "fix" desirable traits. With line-breeding we breed animals that are related, but we are also routinely introducing genes from other lines into the genetic mix. It takes longer to fix the desirable traits this way, but doing so lowers the risk of those problems associated with repeated in-breeding.

Loose line-breeding over successive generations will result in more variations of physical appearance than would in-breeding or tight line-breeding, but will keep the physical look and structure within the same general size and shape, it also carries fewer long term risks.

According to geneticists: Line-breeding can be carried on for many many generations without deleterious effects on the line or breed as long as the individuals involved have few hidden genetic disorders.

Out-Crossing

Out-crossing in terms of pure-bred dogs is the breeding of unrelated dogs. On a pedigree no names will be repeated within the first 5 generations. This type of breeding has both advantages and disadvantages. Which as it turns out are flip sides of the same argument. The greatest genetic diversity are maintained by out -ccrossing, but this also leads to the least consistency in terms of physical appearance and other traits.

Out-crossing does not guarantee that the animals won't develop genetic disorders, but it does tend to reduce the numbers of affected offspring. Best chance of getting an animal that is less prone to developing a genetic disorder comes more from finding a conscientious breeder that screens their animals for hereditary disorders and breeds for the betterment of the breed. All three methods of breeding have their place in a long term, well thought out breeding programme. Talk to the breeder, ask questions as to what their goal is in doing a particular breeding. Ask about the risks and what problems are known to that line. And all lines have some because the perfect dog and the perfect lines are still goals of the future for all breeders.

COW BREEDING

Different Breeds for Different Needs

Dairy cattle are breeds that have been developed to produce milk whereas beef cattle are bred and selected primarily for the production of meat. Dual-purpose breeds do exist and they have been selected for both meat and milk production. In the last 20 years the genetic make up of the UK dairy herd has changed and is still changing. Cows are larger, earlier maturing, are producing far more milk and are more efficient at converting feed into milk.

As it is the birth of the calf that causes milk production, it is critical for the farmer that the cow becomes pregnant again during lactation and produces the next calf one year after the previous one. The average length of productive life of a dairy cow is between four and five lactations, although this varies from one or two to over ten. Since the fifties artificial insemination (AI) has increasingly replaced natural service. Embryo transplants (ET) are also being developed. A surrogate mother may have difficulty delivering, since the foreign calf is often larger than nature intended. When this happens, a rope or chain is attached to the calf's forelegs, and, with a winch-like device the newborn calf is pulled from the womb. Unnecessary suffering for both cow and calf.

Regular Calving

A high annual milk production per cow is dependent on regular calving at about 12- month intervals. Poor reproductive efficiency leads to longer calving indices and a reduction in milk sales per cow per annum. It also leads to an increase in the number of cows culled for failing to conceive. On average, in the UK, only about 85 cows calve per annum for every 100 cows available for breeding. Over thirty-five per cent of all cows culled are disposed of because of reproductive disorders.

Obtaining Semen

Photographer Henry Heap visited the Milk Marketing Board stud bull operation in Mid Wales. He provided Turning Point magazine with a complete set of photographs and commentary of his visit. "A large, handsome, tan bull stood stock still, tethered by the ring tethered through his nose to what looked like a giant circular clothes line or some grotesque merry-go-round. He was kept so motionless by a rubber hood that covered his eyes, which had a similar effect to that of hooding a falcon. A worker commented that the bull would stay still all day if they left the blindfold on. However, this bull was not the source of the semen, but instead what is known as a 'teaser'. The stud bulls are led out by the nose one at a time to mount the motionless hooded bull. After they have mounted the teaser their penis is diverted into a leather vagina which is used to collect the semen. As soon as the process is complete, another bull is led out. The hood keeps the teaser bull stationery for each successive beast to mount. These stud bulls never even see females.

During embryo transfer an embryo can be flushed from the uterus of a genetically superior cow and implanted in a receiver cow of so-called 'poorer quality' for foetal development. Embryos are also taken from dead cows at the slaughterhouse. Hormone treatment is used to induce multiple ovulations and the cow is then inseminated. A few days later several embryos can be flushed from her. For both embryo collection and implanting the cow receives a local anaesthetic injection in the tail. Embryos of the desired breed can then be used. ET maximises the number of offspring from selected breeding females and maximises profit. Do-it-yourself embryo transfer is now taking place in the UK. Courses are run to train in direct embryo transfer techniques using the quick-thaw method for frozen embryos. Although vets are not required to be present at the time of transfer, they must check each animal is suitable for implantation. Farmers genetically improve the herd by using techniques such as MOET (multiple ovulation embryo transfer) which involves inducing a genetically 'superior' cow to produce up to 20 embryos rather than one.

For ease of management, farmers prefer cows to calf at the same time. They, therefore, need to detect when cows come into oestrous. The oestrous detection rate (ODR) is the number of cows observed in oestrus during a 3 week period, in relation to the number of cows expected to be observed. The average ODR on farms is about 60 per cent. One alternative is to synchronise the oestrous period of several animals by injections of prostaglandin or administration of progesterone as an impregnated intraveginal coil implanted for 9-12 days.

ARTIFICIAL INSEMINATION

Artificial insemination (AI) now accounts for around 75 per cent of British cattle births. The cow is tied up and one hand of the inseminator manipulates the cervix through the rectum wall whilst the other discharges semen into the vagina and cervix using an inseminating gun. A regional manager with AI firm Genus the problems associated with the procedure. "Placing a quarter of a cc of diluted semen into an area as small as a penny coin, when the target area is inside 500kg of unco-operative cow, requires a high degree of expertise. The more the cow is able to move about, the harder it becomes to hit the target.

Poor handling facilities cause stress, the cow, the inseminator and both our bank balances." There are many invasive procedures carried out on farm animals that the general public rarely witness. One of many includes making sure the cow is in calf. This is carried out using an ultrasound device much like a small TV set with a long thick wire attachment. A vet will insert their hand and arms length into the vagina of the cow.

The animal is held in a holding device to keep her still. The screen shows the small calf in the middle of a dark shaped uterus. Herds usually calve at the same time (due to human intervention) so all the cows on the farm will usually be checked at the same time.

Jersey Cattle

A small, honey-brown breed of dairy cattle, the Jersey is renowned for the high butterfat content of its milk, as well as a genial disposition. The Jersey cow is quite small, ranging from only 800 to 1200 pounds (360 to 540 kg).

The main factor contributing to the popularity of the breed has been their greater economy of production, due to:

- The ability to carry a larger number of effective milking cows per unit area due to lower body weight, hence lower maintenance requirements
- High butterfat conditions, and to thrive on locally produced food.

Bulls are also small, ranging from 1200 to 1800 pounds (540 to 820 kg), and are notoriously very aggressive. Castrated males can be trained into fine oxen which, due to their small size and gentle nature make them popular with young teamsters. Jersey oxen are not as strong as larger breeds however and are generally out of favour among competitive teamsters. Due to the small size, curious docile character and attractive features of the Jersey cow, small herds were imported into England by aristocratic landowners as props for aesthetic landscapes. As its name implies, the Jersey was bred on the British Channel Island of Jersey.

It apparently descended from cattle stock brought over from the nearby Norman mainland, and was first recorded as a separate breed around 1700. Frank Falle has speculated (on the basis of DNA evidence) that the oldest settlers in Jersey were Danish vikings who had been to Nantes with Hatain, where it is recorded that tribute was given by the King of France to Hatain of 500 cattle to leave that area, whereupon they settled in the Normandy area. The marked resemblance between Jersey cattle and the Nantaise cow would seem to bear this out. While the breed is isolated from outside influence today, this was not always the case. Before 1789 cows would be given as dowry for inter-island marriages between Jersey and Guernsey. This was, however, not widespread. Since 1789, imports of foreign cattle into Jersey have been forbidden by law to maintain the purity of the breed, although exports of cattle and semen have been an important economic resource for the island.

The restriction on the import of cattle was initially introduced in 1789 to prevent a collapse in the export price. The United Kingdom levied no import duty on cattle imported from Jersey. Cattle were being shipped from France to Jersey and then being shipped onward to England to circumvent the tariff on French cattle. The increase in the supply of cattle, sometimes of inferior quality, was bringing the price down and damaging the reputation of Jersey cattle. The import ban stabilised the price and enabled a more scientifically controlled programme of breeding to be undertaken.

Sheep Breeds

The domestic sheep is a multi-purpose animal, and the more than 200 breeds now in existence were created to serve these diverse purposes. Some sources give a count of a thousand or more breeds, but concrete documentation has not been produced for such a number. Almost all sheep are classified as being best suited to furnishing a certain product: wool, meat, milk, hides or a combination thereof (i.e. dual-purpose breeds). Other features that are important when classifying sheep include face colour (generally white or black), the length of the tail, the presence or lack of horns, and the topography for which the breed has been developed.

This last point is especially stressed in the United Kingdom, where breeds are described as either upland (hill or mountain) or lowland breeds. A sheep may also be of a fat-tailed breed, which is a type of dual-purpose sheep common in Africa and Asia with larger deposits of fat within its tail, or the meat from such a breed. Ovine breeds are also grouped based on how well they are suited to producing a certain type of breeding stock. Generally, sheep are thought to be either "ewe breeds" or "ram breeds". Ewe breeds are those that are hardy, and have good reproductive and mothering capabilities—they are for replacing breeding ewes in standing flocks. Ram breeds are selected for rate of weight gain and carcass quality, and are mated with ewe breeds to produce meat lambs. Many breeds, especially rare or primitive ones, fall into neither category strictly. Lowland and upland breeds are also crossed in this fashion, with the hardy hill ewes crossed with larger, fast-growing lowland rams to produce a type of ewe called a mule, which can then be crossed with meat-type rams to produce prime market lambs.

Sheep breeds are categorized by the type of wool they exhibit, irrespective of whether they are favoured for meat or wool production. Fine wool breeds are those that have wool of great crimp and density, and are consequently preferred by the textile industry. Most of these breeds were derived from Merino sheep, and Merinos continue to dominate the world sheep industry. The record for the most valuable sheep ever belongs to an Australian Merino ram which was sold for $16,000 AUD.

Medium wool breeds have wool between the extremes, and are typically fast-growing meat and ram breeds with dark faces. Some major medium wool breeds, such as the Corriedale, are dual-purpose crosses of long and fine-wooled breeds created for high-production commercial flocks. Long wool breeds are the largest of sheep, with lengthy wool and a slow rate of maturation. Long wool sheep are most valued by modern agriculture for crossbreeding to improve the attributes of other sheep types. The American Columbia breed was developed through crossing Lincoln rams—a long wool breed—with fine-wooled Rambouillet ewes.

Coarse or carpet wool sheep are those with a medium to long length wool of characteristic coarseness. Breeds traditionally used for carpet wool show more variability in general traits than other types, the chief requirement being a wool that will not break down under heavy use (as would that of the finer breeds). As the demand for carpet-quality wool declines, some breeders of this type of sheep are attempting to reposition a few of the traditional breeds to alternative purposes. Others have always been primarily meat-class sheep.

A last group of sheep breeds is that of fur or hair sheep, which do not grow wool at all. Hair sheep are similar to the very early type of domesticated sheep kept before wooly breeds were developed, and are primarily raised for meat and pelts. Some modern breeds of hair sheep, such as the Dorper, are the result of crosses between wooled and hair breeds. For producers uninterested in selling wool, hair sheep are highly advantageous, as those who keep hair sheep need not go to the expense of shearing. Hair sheep are also resistant to parasites and hot weather, and are consequently increasing in popularity among a subset of producers.

With the modern rise of corporate agrobusiness and the decline of localized family farms, many breeds of sheep are in danger of extinction. Preference for only those breeds with uniform characteristics and fast growth rates has pushed heritage (or heirloom) breeds to the margins of the sheep industry. Those that remain are maintained through the efforts of conservation organizations, breed registries, and individual farmers dedicated to their preservation.

GENETIC IMPROVEMENT OF DAIRY CATTLE

Dairymen are constantly concerned with improvement of their herds. Dairy cattle improvement is expressed most frequently as increased milk production and less often in terms of changes in body conformation or "type." Improvement in milk production can be considered as being environmental or genetic. Both environmental and genetic factors contribute to variation in milk production. There are environmental and genetic limitations of production. More importantly, management of environment and genetic factors can result in improved performance, *i.e.*, greater milk production.

Genetic Principles of Repeatability

Understanding the concept and knowing the repeatability of economically important traits help considerably in deciding which cows to cull. Repeatability also is important in sire proofs. All repeatability estimates represent some type of correlation. The trait one is considering, and the times and places it was measured, influence the estimate. The correlation between 305-day mature equivalent milk records of first-calf heifers, and their records as second-calf heifers, is about 0.5.

So milk records are said to have a repeatability of 0.5. Conception rate from year-to-year with the same cow has a repeatability of about zero, indicating that there is little or no relationship (correlation) between the number of services it takes for a cow to become pregnant each time. These estimates are useful in making culling and management decisions, since they help us make the best possible guess at how well the cow will do next year if we know how well she did this year. For milk yield, 50 per cent of the apparent superiority (or inferiority) of the cow for this year will appear next year on the average.

If a cow gave 1000 pounds more than her herdmates this year, the best prediction of her yield next year will be 500 pounds more. Naturally, what she actually does is another story; she could be much better or much worse than our guess. The same process is used for below average cows; only one-half of their apparent inferiority will appear the following year on the average. Repeatability is included in the calculation of estimated relative producing ability in DHI. Only a fraction of the cow's performance is permanent (repeatable). Superior cows may do even better next year, but on the average they do worse; inferior cows on the average do better.

Genetic Principle of Heritability

Heritability is a measure of the degree to which a trait such as milk yield is genetically determined. Although not all of the genetic contribution is included, essentially all which is available to a breeder who is trying to select within a breed is included. Obviously heritability is an important factor among the several factors determining how much genetic improvement can be made in any characteristic. One of the disadvantages with the dairy cow, in attempts to make genetic progress, is that most of the economically important traits can be measured only in the female. Another disadvantage is that heritabilities are low or close to zero in several cases. A zero heritability would mean that no genetic change could be expected from usual selection techniques.

A practical idea of the meaning of heritability can be obtained easily. A genetically average group of males is mated to a large group of high producing first calf heifers. These heifers average 1000 pounds of milk more than their contemporaries. Assume also that no extraordinary changes in environment have occurred by the time that daughters from these matings freshen and complete their first record.

How much milk will the daughters produce compared to their herdmates? Answer: about 100 to 150 pounds more. Reason: milk yield has a heritability of about 20 to 30 per cent, and since the sires used were genetically average, and sires and dams contribute about equally to the genotype of their offspring, only 10 to 15 per cent of the superiority of the dams will be seen in the offspring.

A good practical definition of heritability is, therefore, the fraction of the merit (or lack of merit) observed in the parents which can be seen in the offspring, with merit being defined as a deviation from the population average. Note that several reproductive traits have very low heritabilities. Milk yield would be said to be moderately heritable; fat and protein percentages are highly heritable.

One possibility exists for genetic improvement of important traits with zero or very low heritability. This revolves around what geneticists call nonadditive genetic variance. If there is appreciable nonadditive genetic variance, it can be taken advantage of by crossbreeding. Heterosis (hybrid vigour) is a reflection of this type of genetic variability. Unfortunately, although there is some heterosis in reproductive performance, longevity, livability and milk yield, it is not a large amount. Crossbreeding has several disadvantages for dairy cattle, although this discussion cannot be covered here. Few people are ready to advise taking advantage of nonadditive genetic variance in dairy cattle at the present time, and few cattleman are attempting to do so. Reproductive efficiency can be improved considerably by improving management practices while selection is being practiced for more highly heritable traits.

SELECTION AND EXPECTED RESPONSE

Traits which are heritable presumably can be changed by selection, but the degree of heritability is only one of three factors which determine the progress which can be made. The other two are the culling level (intensity of selection) and the variability of the trait. These are logical since the more severely we can cull, the more progress we can make. Also, the more variable a trait is, the higher the population average will be for the selected individuals for any given culling percentage.

It turns out that the amount of progress which can be made in one generation is the product of intensity of selection, heritability for that trait, and the observed variability. Note that if any of the three components is zero, selection progress will be zero also. Good estimates are available for how much progress can be made from selection for most of the economically important characteristics of dairy cattle. Of course, there are several traits which are important, so there is a need to look further than anticipated results from selection for a single trait.

Correlated Response

Knowing what the genetic relationships are among the important traits, plus a few other things, permits us to predict what will happen to a trait such as fat percentage if we ignore it completely and select for milk yield, or vice versa. Though these results pertain to Holsteins, they are similar to results for other breeds. The changes expected in the trait being selected for; values off the diagonal are changes expected to happen in other traits. Heritabilities and variabilities used in these calculations agreed with those generally accepted by dairy cattle geneticists. Intensity of selection in the estimations was fairly conservative. If more severe culling were practiced, greater progress would be made. If milk yield alone is selected for at a reasonable level of intensity, an increase is expected in genetic merit in the next generation of 607 pounds. Since milk yield, fat yield and protein yield are genetically fairly closely related (that is, their genetic correlations are high and positive), an increase is expected in fat and protein yields, even though these were ignored in the selection programme.

Because of the negative genetic correlations between milk yield and fat percentage, and milk yield and protein percentage, these two percentages are expected to go down. Selecting for milk yield and ignoring the other four traits would be expected to result in an increase per generation of 607 pounds of milk, 23.2 pounds of fat and 13.7 pounds of protein, but decreases of 0.036 per cent fat and 0.018 per cent protein. If one attempted to increase fat percentage genetically in his herd by selecting for it and ignoring other traits, he could increase fat percentage. A reasonable increase per generation to be expected is 0.190 per cent. Along with this, however, there would be an actual decrease in milk yield of 287 pounds, although fat yield would increase 24.1 pounds.

Present pricing systems for milk in the United States are based mostly on its weight and fat percentage, although additional factors such as protein percentage may be included also. So the dairyman needs to decide what combination of milk yield and fat percentage and other changes is most profitable for him. He also needs to produce milk that passes minimum legal standards; this is an important factor in herds which already are near the minimum. Beyond this, there are a number of other important traits which must be considered. Dairymen, of course, mentally put all of these factors together in their selection programmes and, depending on their present herd status and present milk pricing, give each trait a certain amount of emphasis. In other words, they weight each characteristic they think is important. Geneticists do the same thing and attempt to establish numerical values for the weights; a numerical index, called a selection index, then can be formulated for each economic situation.

Milk Yield and Composition

Just about every index formulated for the United States

gives yield more emphasis than any other trait, with enough emphasis placed on fat percentage to meet legal standards. Protein has become more important in pricing systems in recent years.

Reproductive Performance

Efficient reproduction is very important economically because a cow must first reproduce to produce. However, most measures of heritability are zero or close to zero, suggesting that little or no improvement can be made by selection. There already is considerable selection pressure being exerted on improving reproductive performance without any conscious effort being made by the breeder. This occurs because sterile males and females leave no offspring, and those with low efficiency leave a lot fewer than those of higher reproductive performance. So within a given herd culling for reproductive failure would have little effect on the genetic merit for reproduction of the herd. Now that large numbers of records are being obtained on reproduction of daughters of AI sires, however, it may turn out that this information can be taken advantage of in selecting which proven AI sires to use in the herd. Even though heritability of most single measures of reproduction are close to zero, estimates of average genetic merit of a large group of daughters for reproductive efficiency are much more reliable.

Desirable Body Type

Desirable type has appealed to dairymen for centuries, and major changes in appearance in dairy cattle have occurred over the years, accompanied by increases in milk yield. This fact does not itself mean that the two traits have (or ever have had) an appreciable, positive genetic correlation. Regardless of what has gone before, the breeder must concern himself with the present facts of life in his selection programme. The genetic relationship between type and production seems close to zero and may even be negative. There have been a number of changes in the breed association type classification programmes and policies in recent years, and evaluation of all such changes cannot yet be made. A number of scoring systems also have been developed by others such as universities and AI studs. Yet the final value of desirable type must be in its relationship with production and longevity of the cow.

Except for the breeder who derives an important fraction of his income from sale of breeding stock, dairymen should realise (according to research) that only 2 to 6 per cent of their income is related to variability in body confirmation. USDA researchers estimated several years ago that giving equal emphasis to type and production would reduce the possible increase in milk yield by 29 per cent. Though perhaps the final judgment on emphasis to be placed on desirable type cannot yet be made, it appears that if major attempts are made to improve type, they will be made at the sacrifice of possible genetic increases in milk yield.

Longevity

Longevity may be defined as the ability to avoid death or culling, the former being mostly for disease or accident and the latter for reproductive failure, low milk yield, mastitis, etc. Few cows die simply from old age. Of the various possible reasons for loss, some have a genetic basis but few are highly heritable. In any event, high producing cows remain in the herd longer than low producers. Natural selection pressures also work in favour of the ability to resist death or culling. Selection for high milk yield, plus natural selection, seem to be exerting sufficient selection pressure for increased longevity such that the breeder need not give it major emphasis in his programme.

Body Size and Weight

Within each dairy breed, large cows give more milk, on the average, than small cows, but they also require more feed. Research suggests that the additional costs involved with larger cows about equal the additional income derived from them. In the case of two cows of equal production but different size, one actually would expect the smaller cow to be more profitable. Most measures of size (weight, height, etc.) show appreciable heritability and variability, so larger cows could doubtless be developed if desired.

However, most research now suggests that milk yield in dairy cows can be increased considerably without there being any increase in size. Selection for increased body size could be successful, but the correlated response in milk yield though probably positive would be close to zero. Though North Carolina researchers indicated that it might still be wise to check cows considerably above or below the breed average in body weight for productivity and longevity, they concluded that "undue emphasis on larger cows in the short run could lead to the development of less profitable animals."

Other Traits

Many others are economically important and are heritable to some degree. Many qualitative (simply inherited) traits, ranging from lethals on one extreme to aesthetic traits on the other, can be important under some circumstances. Artificial insemination, particularly frozen semen, has caused concern because of possible widespread dissemination of undesirable genes. The genetics of the situation have been studied, however, and with present knowledge potential dangers apparently can be avoided. Traits other than those mentioned also are important. Mastitis resistance, is of extreme economic value and has measurable heritability. A certain amount of direct pressure for resistance will be exerted from culling of mastitic animals, although clinical cases may remain in a dairy herd for years under proper treatment.

Present knowledge of genetics of dairy cattle suggests that major selection emphasis should be placed on milk yield, with enough selection pressure on fat percentage to maintain composition of milk at legal standards. Other components of milk may become important if pricing systems change to emphasize them. Care should be taken to include as few traits as practical in a selection programme, since progress in milk yield may be sacrificed to a degree such that total profit per cow may decrease. This is especially true with traits of low heritability and only slight genetic relationship to milk yield. Reproductive efficiency and resistance to mastitis may appear as economically important traits in future sire proving systems.

The Animal Model

The Animal Model does many things notably different from previous evaluation systems. These include simultaneous evaluation of all sires and cows. This means all information on all relatives is used to increase the accuracy of any one animal's evaluation. This also means that cow families are directly included in the evaluation of all sires and cows, effectively eliminating the basis for previous objections by some breeders. All female ancestors and descendants are used in the evaluation of both sires and cows, weighted by how closely they are related. Since all known relatives of both registered and grade cows are used, computing the evaluations for the Holstein breed requires simultaneous solution of about 50 million equations. Other changes include correction for genetic merit of mates, use of a new genetic base, new ways of using records on cows missing first records and cows changing herds, and definition of different management groups for registered and grade cows.

There is some new terminology with the Animal Model emphasizing that new methods are truly different. Predicted Difference (PD) changes to Predicted Transmitting Ability (PTA), or one-half the breeding value. This is the prediction of what the sire transmits to his offspring. Cow Index changes to Predicted Transmitting Ability (PTA).

Repeatability (RPT) changes to reliability. Repeatability always has been a misnomer as used by the USDA in this context anyway. Since reliability includes contributions from additional relatives, reliability figures should be higher than the old repeatabilities. The biggest difference is for sires of cows with few daughters compared to those with many. A sire with no daughters would have a reliability of 37 per cent. Percentile is sire's ranking based on the active AI bull population; for cows it will be based on cows with recent lactations. The rankings are based on an index including milk, fat, and protein evaluations. Previously they were based on an index including only milk and fat. As good as the new Animal Model system is, it cannot correct for preferential treatment of daughters of certain sires. No genetic evaluation system can. Random AI sampling of young sires and wide daughter distribution will continue to be important in establishing accurate first evaluations. This also is true for sires with relatively few daughters.

Economic Evaluation of Sires and Their Semen

Predicted transmitting ability for dollar value (PTA PRO$) is based on milk, fat, and protein evaluations. Current national prices are used. For the July 1989 run these were $11.20 per hundredweight of milk containing 3.5 per cent milk fat and 3.2 per cent protein. Hence the equation for estimating PTA PRO$ is PTA PRO$ = $.0242 (PTA milk) + $ 1.54 (PTA fat) + $ 1.09 (PTA protein). Note that the dollar value is computed from pounds fat and protein produced, not percentages. Percentile rankings are based on this index.

In Florida, the milk price is higher than almost everywhere else in the United States, and the dollar values of superior sires should be proportionally higher. The ranking of bulls by evaluations would not change much, but the net value over their semen price would. Taking advantage of considerable previous research, Marsha Wilcox and coworkers developed techniques to aid dairymen in the selection of semen to purchase for herd use. The concept is that of net present value (discounted net income) of semen. Although the method is computationally elaborate, availability of microcomputers to dairymen will permit each to develop a system of evaluating semen based on a set of selection criteria tailored to individual needs. Information necessary for semen selection include sire proof for milk, fat, protein, type (PDT), interest rate, and such management parameters for the individual herd as conception rate, calving interval and female mortality rate. Outcome of this linear net merit index for semen is Present Value Dollars (PV$). These researchers then evaluated a number of possible economic and management situations to determine the effects on sire selection; considerable shifting of rank occurred with changes in criteria. For one sample selection programme (all selection pressure on milk, none on desirable type), rank correlations of PV$ and PDM, and PV$ and PD$ + semen price, were.56 and.52. Such procedures should prove useful and profitable to dairymen who are able to characterize their own management parameters and to define their selection goals. Thus, the concepts of proving sires are not difficult. Efforts to obtain better estimates, thus increasing rates of genetic improvement, continue to make the exact calculations and formulas very complex.

SELECTING SIRES FOR A SINGLE HERD

Most of the factors involved in sire selection have been covered in earlier sections, but one major question remains. Will the sire which has been proven to be superior in other herds also be superior in my herd? Older breeders will remember the term "nicking," which is not too often heard anymore. If a sire did well in one herd and poorly in a second, he was said to nick well in the first, poorly in the second.

Today, researchers look at the importance of nicking by studying the importance of sire by farm interactions. Such interactions arise if there are real changes in the relative performance of sires (usually measured by the milk yield of their daughters) as they are used on various farms. Should there be an appreciable number of sires which rank high in some herds or farms, and low in others, and vice versa for other sires, this would show up as a sire-by-herd interaction.

The importance of this question is apparent; if there were such interactions, a dairymen would be unable to rely on USDA sire proofs, since he could not depend on sires with superior proofs to be superior in his herd. This was an area that obviously needed research, and during the past several years considerable research has been published on the subject. A classic study by North Carolina scientists involved sire-by-region interactions. The two regions involved were North Central and Southeastern United States and the question essentially was, "Do AI sires used in these two areas rank about the same, or are there some who are outstanding in one region but poor in the other, and vice versa?" Much of the semen used by Florida dairymen is from sires bred, house and proven in the Eastern and North Central United States, so this question was of immediate concern.

Results showed that this particular type of interaction was not an important factor, and indeed was almost nonexistent. The authors stated, "provided adequate unbiased progeny information is utilized, a bull may be safely selected and used for artificial insemination in one region, even though he was evaluated in the other region." Studies on the importance of sire by herd interactions have been in very close agreement, with one being conducted by Florida researchers on records of Florida DHIA Guernsey, Holstein and Jersey first-calf heifers. Everyone recognizes that even the best bull may have a certain number of poor daughters and that a dairyman who has only a few daughters of a particular sire may have been unlucky, by chance, and gotten one or more of them. The opposite can happen also with poor sires.

The problem is to determine how important the interactions are overall, realizing that a certain amount of shifting in rankings of sires can occur by chance, particularly if a sire is represented in one or more herds by only a small number of daughters. The research found that under Florida DHIA herd conditions; these interactions were of little or no practical or theoretical consequence for milk and fat yields. There was some evidence among Guernsey's and Jerseys of an interaction for fat percentage; this has been suggested in other research studies also, but many of the scientists who have studied interactions have not looked at fat percentage. In summary, it appears that sires proven superior for milk and fat yields in other herds can be used with confidence in Florida dairy herds, without concern for possible sire by herd interactions. The possibility exists for the presence of interactions with fat percentage.

FINDING SUPERIOR FEMALES

Measurements for most of the economically important

traits can be made directly on females. It frequently takes a long time to get the measurements of interest, however. Mature body size, would have to wait until the cow is 6 to 8 years old. An estimate of milk yield must wait until the first lactation is completed (at about 3 years of age), unless one accepts an estimate of the lactation based on the first several months.

These points are mentioned because genetic progress is influenced by the time at which measurements are made, and their reliability. Take milk yield as an example. An estimate of the milk producing ability of a cow comes with her first milk record. Certainly, however, one knows more about her when there are two records rather than one, and even more with three records. Generally, the more records available, the more reliable the estimate of her real ability; but it takes longer to get more records. An estimate of the producing ability of a cow was called "Most Probable Producing Ability" years ago by Dr. J.L. Lush, well known animal geneticist. This measure takes into consideration the idea of repeatability, and shows how the number of records a cow has should influence the estimate of her producing ability. If a heifer produces more than her herdmates during her first lactation, we will certainly guess that she is a better than average animal.

If the average of her first two records is above herdmates, one would be even more confident that she is truly above average, and her good performance did not just happen from some favourable accident which probably will not happen again. The calculations are simple and can be applied to any trait of interest. Although the concept is rarely mentioned today, since availablity of computers makes more sophisticated procedures feasible, MPPA serves as the basis for modern methods. State the milk record, or any other measurement, as a deviation from herdmates. If the cow has two measurements, state each as a deviation, and take the average. MPPA then can be calculated as:

$$\text{MPPA} = (\text{Average deviation}) \times [(nr)/(1 + nr - r)]$$

where n = the number of measurements and r = the repeatability. How much better two measurements or records are than one, as far as reliability, is determined by a similar equation.

$$\text{Gain in accuracy} = 1 - [(1 + nr - r) / n]$$

with n and r as in MPPA. The more records one has on a cow, the more one can accept her records at face value. If she has six records averaging 2000 pounds better than her herdmates, the best guess of her MPPA is 1714 pounds better.

Which cow should be selected as a dam for a future herd sire, a cow with two records averaging 3000 pounds above herdmates, or one with four records averaging 2400 pounds above? The two estimates of MPPA are close, being 2000 pounds for the cow with two records and 1920 pounds for the older cow. One probably would look at other characteristics of the two animals, particularly fat per cent but possibly others, before making a decision. The more sophisticated and useful application of this concept, of course, is that it is incorporated into calculations of ERPA.

GENETIC CHANGE IN MILK YIELD

The theoretical upper limits of genetic change in milk yield have been known for quite some time. Knowing the various repeatabilities, heritabilities, generation intervals and selection pressures (intensities), one can calculate what would be expected to happen. This tells nothing about what is actually going on, of course, but it does tell what one can expect under the best conditions.

Potential

If milk yield is given maximum emphasis in the selection programme, with little or no pressure being exerted on other traits, and efficient use is made of an AI sire proving programme, the maximum attainable genetic change in milk yield is slightly over 2.0 per cent per year.

Most of this, over 90 per cent, theoretically would come directly or indirectly from sire selection. In a Holstein population averaging 15,000 pounds per cow, this would amount to 300 pounds per year. Though this estimate has appeared disappointingly small to some breeders, it is more than the actual total change which has been attained in the United States.

Observed

It is only in recent years that it has been possible to obtain reliable estimates of the actual genetic change which has occurred in milk production. A major problem in the actual measurement of such progress is that environmental and genetic changes are confounded in such a manner over time as to make their separation extremely difficult. With the availability of high-speed large-capacity electronic computers, many of the problems can now be handled. Use of the animal model has improved the accuracy of the estimates. As far as progress arising from selection of males and females, theoretical estimates has been right on target.

About 90 to 92 per cent of all genetic change in milk yield has occurred from sire selection; this represents the selection of males and females to produce herd sires and the selection of males to produce herd dams. The remaining 8 to 10 per cent of genetic progress arises from selection of females to produce herd dams. The fact that we must raise nearly all female calves born alive in a population to maintain population size, and the difficulties of evaluating the true genetic merit of dams, precludes much selection pressure from selection of dams. Average annual genetic change in milk yield of tested dairy cattle has been variable but seems to be about 0.7 per cent of their average levels. This would amount to about 105 pounds per cow in a 15,000 pound herd, or 140 pounds in a 20,000 pound herd. In individual herds, one would expect to find genetic changes which were considerably higher or lower. If 2.0 per cent annual genetic change is possible, as theory suggests, dairymen are progressing genetically only about one-third as fast as possible.

DOG

The dog is a domesticated subspecies of the wolf, a mammal of the Canidae family of the order Carnivora. The term encompasses both feral and pet varieties and is also sometimes used to describe wild canids of other subspecies or species. The domestic dog has been one of the most widely-kept working and companion animals in human history, as well as being a food source in some cultures. There are estimated to be 400,000,000 dogs in the world. The dog has developed into hundreds of varied breeds. Height measured to the withers ranges from a few inches in the Chihuahua to a few feet in the Irish Wolfhound; colour varies from white through grays to black, and browns from light to dark in a wide variation of patterns; and, coats can be very short to many centimeters long, from coarse hair to something akin to wool, straight or curly, or smooth. In breeding circles, a male canine is referred to as a *dog,* while a female canine is called a *bitch.* The father of a litter is called the *sire,* and the mother of a litter is called the *dam.* Offspring are generally called *pups* or *puppies* until they are about a year old. A group of offspring is a *litter.* The process of birth is *whelping*. Many terms are used for dogs that are not purebred.

Origin and Evolution

Based on DNA evidence, the wolf ancestors of modern dogs diverged from other wolves about 100,000 years ago, and dogs were domesticated from those wolf ancestors about 15,000 years ago. This date would make dogs the first species to be domesticated by humans. Evidence suggests that dogs were first domesticated in East Asia, possibly China, and some of the peoples who entered North America took dogs with them from Asia. As humans migrated around the planet a variety of dog forms migrated with them.

The agricultural revolution and subsequent urban revolution led to an increase in the dog population and a demand for specialization. These circumstances would provide the opportunity for selective breeding to create specialized working dogs and pets. This ancient mosaic, likely Roman, shows a large dog with a collar hunting a lion. Molecular systematics indicate that the domestic dog descends from one or more populations of wild wolves. As reflected in the nomenclature, dogs are descended from the wolf and are able to interbreed with wolves. The relationship between human and canine has deep roots. Converging archaeological and genetic evidence indicate a time of domestication in the late

Upper Paleolithic close to the Pleistocene/Holocene boundary, between 17,000 and 14,000 years ago. Fossil bone morphologies and genetic analysis of current and ancient dog and wolf populations have not yet been able to conclusively determine whether all dogs descend from a single domestication event, or whether dogs were domesticated independently in more than one location.

Domesticated dogs may have interbred with local populations of wild wolves on several occasions (a process known in genetics as introgression). The earliest dog fossils, two crania from Russia and a mandible from Germany, date from 13,000 to 17,000 years ago. Their likely ancestor is the large northern Holarctic wolf, *Canis lupus lupus*. Remains of smaller dogs from Mesolithic cave deposits in the Middle East, dated to around 12,000 years ago, have been interpreted as descendants of a lighter Southwest Asian wolf, *Canis lupus arabs*.

Rock art and skeletal remains indicate that by 14,000 years ago, dogs were present from North Africa across Eurasia to North America. Dog burials at the Mesolithic cemetery of Svaerdborg in Denmark suggest that in ancient Europe dogs were valued companions. Genetic analyses have so far yielded divergent results. The ancestors of dogs split off from other wolves between 75,000 and 135,000 years ago, while a subsequent analysis indicated a "common origin from a single gene pool for all dog populations" between 40,000 and 15,000 years ago in East Asia. Both sets of dates must be reevaluated in light of recent findings showing that poorly calibrated molecular clocks have systematically overestimated the age of geologically recent events. On balance, and in agreement with the archaeological evidence, 15,000 years ago is the most likely time for the wolf-dog divergence.

Development of Dog Breeds

Dogs have been bred into a variety of shapes, colours and sizes. Variation can be wide even within a breed, as with these Cavalier King Charles Spaniels. There are numerous dog breeds, with over 800 being recognized by various kennel clubs worldwide. Many dogs, especially outside the United States of America and Western Europe, belong to no recognized breed. A few basic breed types have evolved gradually during the domesticated dog's relationship with humans over the last 10,000 or more years, but all modern breeds are of relatively recent derivation. Many of these are the product of a deliberate process of artificial selection. Because of this, some breeds are highly specialized, and there is extraordinary morphological diversity across different breeds. Despite these differences, dogs are able to distinguish dogs from other kinds of animal.

The definition of a dog breed is a matter of some controversy. Depending on the size of the original founding population, closed gene pool breeds can have problems with inbreeding, specifically due to the founder effect. Dog breeders are increasingly aware of the importance of population genetics and of maintaining diverse gene pools. Health testing and new DNA tests can help avoid problems, by providing a replacement for natural selection. Without selection, inbreeding and closed gene pools can increase the risk of severe health or behavioural problems. Some organizations define a breed more loosely, such that an individual may be considered of one breed as long as 75 per cent of its parentage is of that breed. These considerations affect both pets and the show dogs entered in dog shows.

Even prize-winning purebred dogs sometimes possess crippling genetic defects due to founder effect or inbreeding. These problems are not limited to purebred dogs and can affect cross-breed populations. The behaviour and appearance of a dog of a particular breed can be predicted to a degree, while mixed-breed dogs show a broader range of innovative appearance and behaviour. Mixed-breed dogs or Mongrels (also called "mutts") are dogs that do not belong to specific breeds, being mixtures more than two in variant percentages. Mixed breed dogs and purebred dogs are both suitable as companions, pets, working dogs, or competitors in dog sports. Sometimes different breed dogs are deliberately bred, to create cross-breeds such as the Cockapoo, a mixture of Cocker Spaniel and Miniature Poodle. Such deliberate crosses may display some degree of hybrid vigour and other desirable traits, but may or may not inherit any of the desired traits of their parents, such as temperament or a particular colour or coat. Without genetic testing of the parents, the crosses can end up inheriting genetic defects that occur in both parental breeds. A breed is a group of animals that possesses a set of inherited characteristics that distinguishes it from other animals within the same species. Deliberately crossing two or more breeds is also a manner of establishing new breeds, but it is only a breed when offspring will reliably demonstrate that particular set of characteristics and qualities.

Breed Popularity

Breed popularity varies widely over time and in different parts of the world and different segments of the population. Counting by American Kennel Club (AKC) registration (not by licensing registration or by United Kennel Club (UKC) registration, which could present different statistics), the Labrador Retriever has been the United States's most commonly registered breed of dog since 1991. However, even within parts of the United States, popularity varies; for example, in 2005 the most-registered breed in New York City was the Poodle while the Yorkshire Terrier was the second-most-registered breed in Houston.

However, animal shelters in many parts of the United States report that the most-commonly available dog for adoption is the American Pit Bull Terrier or pit bull-type mixes, making up as much as 20 per cent of dogs available for adoption, none of which would be registered with the AKC. Two decades ago, in 1983, the AKC's top two registered breeds were the American Cocker Spaniel and the Poodle.

In the United Kingdom, The Kennel Club reports that the most-registered breed from at least 1999 to 2005 was the Labrador Retriever. It rounds out the top three for 1999 to 2005 with the German Shepherd Dog, also popular in the States, and the English Cocker Spaniel, which is no longer even in the top ten in the States. In the United Kingdom a national dog adoption and rescue service indicates that the most common breed appearing in shelters is the Greyhound followed by the Staffordshire Bull Terrier.

Physical Characteristics

Modern dog breeds show more variation in size, appearance, and behaviour than any other domestic animal. Within the range of extremes, dogs generally share attributes with their wild ancestors, the wolves. Dogs are predators and scavengers, possessing sharp teeth and strong jaws for attacking, holding, and tearing their food. Although selective breeding has changed the appearance of many breeds, all dogs retain basic traits from their distant ancestors. Like many other predatory mammals, the dog has powerful muscles, fused wristbones, a cardiovascular system that supports both sprinting and endurance, and teeth for catching and tearing. Compared to the bone structure of the human foot, technically, dogs walk on their toes.

Differences from Other Canids

Compared to equally sized wolves, dogs tend to have 20 per cent smaller skulls and 10 per cent smaller brains, as well as proportionately smaller teeth than other canid species. Dogs require less calories to function than wolves. Their diet of human refuse in antiquity made the large brains and jaw muscles needed for hunting unnecessary. It is thought by certain experts that the dog's limp ears are a result of atrophy of the jaw muscles. Dogs tend to be poorer than wolves and coyotes at observational learning, being more responsive to instrumental conditioning.

Dogs have thicker skins than similairly sized wolves. Inuits tend to favour dog skin over wolf skin due to the formers resistance to tear. Unlike wolves, but like coyotes, domestic dogs have sweat glands on their pawpads. The paws of a dog are half the size of those of a wolf, and their tails tend to curl upwards, another trait not found in wolves. Feral dogs show little of the complex social structure or dominance heirarchy present in wolf packs. For dogs, other members of their kind are of no help in locating food items, and are more like competitors. Dogs differ from wolves and most other large canid species by the fact that they do not regurgitate food for their young, nor the young of other dogs in the same territory.

Many dogs, such as this American Water Spaniel, have had their natural hunting instincts suppressed or altered to suit human needs. Like most mammals, dogs are dichromats and have colour vision equivalent to red-green colour blindness in humans. Different breeds of dogs have different eye shapes and dimensions, and they also have different retina configurations. Dogs with long noses have a *"visual streak"* which runs across the width of the retina and gives them a very wide field of excellent vision, while those with short noses have an *"area centralis"*—a central patch with up to three times the density of nerve endings as the *visual streak*—giving them detailed sight much more like a human's.

Some breeds, particularly the sighthounds, have a field of vision up to 270° (compared to 180° for humans), although broad-headed breeds with short noses have a much narrower field of vision, as low as 180°. Dogs detect sounds as low as the 16 to 20 Hz frequency range (compared to 20 to 70 Hz for humans) and above 45 kHz (compared to 13 to 20 kHz for humans), and in addition have a degree of ear mobility that helps them to rapidly pinpoint the exact location of a sound. Eighteen or more muscles can tilt, rotate and raise or lower a dog's ear. Additionally, a dog can identify a sound's location much faster than a human can, as well as hear sounds up to four times the distance that humans are able to. Those with more natural ear shapes, like those of wild canids like the fox, generally hear better than those with the floppier ears of many domesticated species.

Dogs have nearly 220 million smell-sensitive cells over an area about the size of a pocket handkerchief (compared to 5 million over an area the size of a postage stamp for humans). Some breeds have been selectively bred for excellence in detecting scents, even compared to their canine brethren. What information a dog actually detects when he is scenting is not perfectly understood; although once a matter of debate, it now seems to be well established that dogs can distinguish two different types of scents when trailing, an air scent from some person or thing that has recently passed by, as well as a ground scent that remains detectable for a much longer period. The characteristics and behaviour of these two types of scent trail would seem, after some thought, to be quite different, the air scent being intermittent but perhaps less obscured by competing scents, whereas the ground scent would be relatively permanent with respect to careful and repetitive search by the dog, but would seem to be much more contaminated with other scents.

In any event, it is established by those who train tracking dogs that it is impossible to teach the dog how to track any better than it does naturally; the object instead is to motivate it properly, and teach it to maintain focus on a single track and ignore any others that might otherwise seem of greater interest to an untrained dog. An intensive search for a scent, for instance searching a ship for contraband, can actually be very fatiguing for a dog, and the dog must be motivated to continue this hard work for a long period of time. Domestic dogs often display the remnants of counter-shading, a common natural camouflage pattern. The general theory of countershading is that an animal that is lit from above will appear lighter on its upper half and darker on its lower half where it will usually be in its own shade. This is a pattern that predators can learn to watch for. A countershaded animal will have dark colouring on its upper surfaces and light colouring below.

This reduces the general visibility of the animal. One reminder of this pattern is that many breeds will have the occasional "blaze", stripe, or "star" of white fur on their chest or undersides. Dogs can generate large amounts of energy for a short period of time. A dog's heart and lungs are oversized relative to its body and its normal everyday needs. A dog also has relatively more red blood cells than a human. Most of the time the dog will keep the extra red blood cells stored in its spleen. When the animal enters into a situation where its full metabolism is required, such as play, catching game, or fighting other dogs, the extra cells are released into the bloodstream. The "oversized" heart and lungs will now be running at full capacity, and the animal will have an enhanced ability to engage in aerobic activity. This activity will produce internal heating. Dogs, being covered in fur, are limited in their ability to cool down. After a short time the animal must either cease its athletic activity or risk harming itself from overheating. One can easily observe this pattern of intense activity followed by rest periods in puppies. During the rest phase the spleen collects red blood cells and the animal may pant to cool down.

Dogs are very social animals, but their personality and behaviour vary with breed as well as how they are treated by their owners and others who come in contact with them. It is not uncommon for dogs to attack humans and other animals; however, this is usually because of lack of care or improper upbringing by its owner. Many dogs can be trained to skillfully perform tasks not natural to canines, such as in this dog agility competition. Dogs are valued for their intelligence. This intelligence is expressed differently with different breeds and individuals, however. For example, Border Collies are noted for their ability to learn commands, while other breeds may not be so motivated towards obedience, but instead show their cleverness in devising ways to steal food or escape from a yard. Being highly adaptable animals themselves, dogs have learned to do many jobs as required by humans over the generations.

Dogs are employed in various roles across the globe, proving invaluable assets in areas such as search-and-rescue; law enforcement (including attack dogs, sniffer dogs and tracking dogs); guards for livestock, people or property; herding; Arctic exploration sled-pullers; guiding the blind and acting as a pair of ears for the deaf; assisting with hunting, and a great many other roles which they may be trained to assume. Most dogs rarely have to deal with complex tasks and are unlikely to learn relatively complicated activities (such as opening doors) unaided. Some dogs (such as guide dogs for the visually impaired) are specially trained to recognize and avoid dangerous situations. The meaning of "intelligence" in general, not only in reference to dogs, is hard to define. Some tests measure problem-solving abilities and others test the ability to learn in comparison to others of the same age. Defining it for dogs is just as difficult. It is likely that dogs do not have the ability to premeditate an action to solve a problem.

For example, the ability to learn quickly could be a sign of intelligence. Conversely it could be interpreted as a sign of a desire to please. In contrast, some dogs who do not learn very quickly may have other talents. An example is breeds that are not particularly interested in pleasing their owners, such as Siberian Huskies. Huskies are often fascinated with the myriad of possibilities for escaping from yards, catching small animals, and often figuring out on their own numerous inventive ways of doing both. Assistance dogs are also required to be obedient at all times. This means they must learn a tremendous number of commands, understand how to act in a large variety of situations, and recognize threats to their human companion, some of which they might never before have encountered. Many owners of livestock guardian breeds believe that breeds like the Great Pyrenees or the Kuvasz are not easily trained because their stubborn nature prevents them from seeing the point of such commands as "sit" or "down".

Hounds may also suffer from this type of ranking. These dogs are bred to have more of a "pack" mentality with other dogs and less reliance on a master's direct commands. While they may not have the same kind of intelligence as a Border Collie, they were not bred to learn and obey commands quickly, but to think for themselves while trailing game. Dogs are highly social animals. This can account for their trainability, playfulness, and ability to fit into human households and social situations. These attributes have earned dogs a unique position in the realm of interspecies relationships despite being one of the most effective, voracious, and potentially dangerous predators.

Dogs and humans at times co-operate in some of the most effective hunting in the animal world; in that context, dogs are superpredators. The loyalty and devotion that dogs demonstrate as part of their natural instincts as pack animals closely mimics the human idea of love and friendship, leading many dog owners to view their pets as full-fledged family members. Conversely, dogs seem to view their human companions as members of their pack, and make few, if any, distinctions between their owners and fellow dogs. Dogs fill a variety of roles in human society and are often trained as working dogs. For dogs that do not have traditional jobs, a wide range of dog sports provide the opportunity to exhibit their natural skills. In many countries, the most common and perhaps most important role of dogs is as companions. Dogs have lived and worked with humans in so many roles that their loyalty has earned them the unique sobriquet "man's best friend". However, some cultures consider dogs to be unclean. In some parts of the world, dogs are raised as livestock to produce dog meat for human consumption. In many places, consumption of dog meat is discouraged by social convention or cultural taboo.

Laughter might not be confined or unique to humans, despite Aristotle's observation that "only the human animal laughs". The differences between chimpanzee and human laughter may be the result of the adaptations that enable human speech. However, some behavioural psychologists argue that self-awareness of one's situation, or the ability to identify with somebody else's predicament, are prerequisites for laughter, so animals are not really laughing in the same way that humans do. The dog-laugh sounds similar to a normal pant. However by analyzing the pant using a spectrograph, this pant varies with bursts of frequencies, resulting in a laugh.

When this recorded dog-laugh vocalization is played to dogs in a shelter setting, it can initiate play, promote pro-social behaviour, and decrease stress levels. In a study by Simonet, Versteeg, and Storie, one hundred and twenty subject dogs residing in a mid-size county animal shelter were observed. Dogs ranging from 4 months to 10 years of age were compared with and without exposure to a dog-laugh recording. The stress behaviours measured included panting, growling, salivating, pacing, barking, cowering, lunging, play-bows, sitting, orienting and lying down. The study resulted in positive findings. Exposure to the dog laughing recording resulted in the following: significantly reduced stress behaviours, increased tail wagging and the display of a play-face when playing was initiated, and more frequent pro-social behaviour such as approaching and lip licking. This research suggests exposure to dog-laugh vocalizations can calm and possibly increase shelter adoptions.

HORSE BREEDING

Horse breeding refers to reproduction in horses, and particularly the human-directed process of selective breeding of animals, particularly purebred horses of a given breed. While feral and wild horses breed successfully without human assistance, planned matings can be used to produce specifically desired characteristics in domesticated horses. Furthermore, modern breeding management and technologies can increase the rate of conception, a healthy pregnancy, and successful foaling. The male parent of a horse, a stallion, is commonly known as the sire and the female parent, the mare, is called the dam. Both are genetically important, as each parent provides half of the genetic makeup of the ensuing offspring, called a foal. (Contrary to popular misuse, the word "colt" refers to a young male horse only; "filly" is a young female.)

Though many amateur horse owners may simply breed a family mare to a local stallion in order to produce a companion animal, most professional breeders use selective breeding to produce individuals of a given phenotype, or breed. Alternatively, a breeder could, using individuals of differing phenotypes, create a new breed with specific characteristics. Some breeders consider the quality of the sire to be more important than the quality of the dam. However, other breeders maintain that the mare is the most important parent. Because stallions can produce far more offspring than mares, a single stallion can have a greater overall impact on a breed. However, the mare may have a greater influence on an individual foal because its physical characteristics influence the developing foal in the womb and the foal also learns habits from its dam when young. Foals may also learn the "language of intimidation and submission" from their dam, and this imprinting may affect the foal's status and rank within the herd. Many times, a mature horse will achieve status in a herd similar to that of its dam; the offspring of dominant mares become dominant themselves.

Estrus Cycle of the Mare

The estrous cycle controls when a mare is sexually receptive towards a stallion, and helps to physically prepare the mare for conception. It generally occurs during the spring and summer months, although some mares may be sexually receptive into the late fall, and is controlled by the photoperiod (length of the day), the cycle first triggered when the days begin to lengthen. The estrus cycle lasts about 19-22 days, with the average being 21 days. As the days shorten, the mare returns to a period when she is not sexually receptive, known as anestrus. Anestrus prevents the mare from conceiving in the winter months, as that would result in her foaling during the harshest part of the year, a time when it would be most difficult for the foal to survive.

This cycle contains 2 phases:

1. *Estrus, or Follicular, phase*: 5-7 days in length, when the mare is sexually recepetive to a stallion. Estrogen is secreted by the follicle. Ovulation occurs in the final 24-48 hours of estrus.
2. *Diestrus, or Luteal, phase*: 14-15 days in length, the mare is not sexually recepetive to the stallion. The corpus luteum secretes progesterone.

Although 16 per cent of mares have double ovulations (allowing them to twin), this does not affect the length of time of estrus or diestrus.

Effects on the Reproductive System During the Estrous Cycle

Changes in hormone levels can have great affects on the physical characteristics of the reproductive organs of the mare, thereby preparing, or preventing, her from conceiving.

- *Uterus*: Increased levels of estrogen cause edema within the uterus, making it feel heavier, and the uterus loses its tone. This edema decreases following ovulation, and the muscular tone increases. High levels of progesterone do not cause edema within the uterus. The uterus becomes flaccid during anestrus.
- *Cervix*: The cervix relaxes right before estrus occurs, and the secretions of the cervix increase. High progesterone levels (during diestrus) cause the cervix to close.
- *Vagina*: The portion of the vagina near the cervix becomes engorged with blood right before estrus. The vagina becomes relaxed and secretions increase.
- *Vulva*: Relaxes right before estrus begins. Becomes dry, and closes tightly, during diestrus.

Hormones Involved in the Estrus Cycle, During Foaling, and after Birth

The cycle is controlled by several hormones which regulate the estrous cycle, the mare's behaviour, and the reproductive system of the mare. The cycle begins when the increased day length causes the pineal gland to reduce the levels of melatonin, thereby allowing the hypothalamus to secrete GnRH.

- *GnRH (Gonadotropin releasing hormone):* Secreted by the hypothalamus, causes the pituitary to release of 2 gonadotrophins: LH and FSH.
- *LH (Luteinizing hormone):* Levels are highest 2 days following ovulation, then slowly decrease over 4-5 days, dipping to their lowest levels 5-16 days after ovulation. Stimulates the follicle to develop, which in turn causes an increase in estrogen. Unlike most mammals, the mare does not have an increase of LH right before ovulation.
- *FSH (Follicle-stimulating hormone):* Secreted by the pituitary, causes the ovarian follicle to develop. Levels of FSH rise slightly at the end of estrus, but have their highest peak about 10 days before the next

ovulation. FSH is inhibited by inhibin, at the same time LH and estrogen levels rise, which prevents immature follicles from continuing their growth.

- *Estrogen:* Secreted by the developing follicle, it causes the pituitary gland to secrete more LH (therefore, these 2 hormones are in a positive feedback loop). Additionally, it causes behavioural changes in the mare, making her more recpetive towards the stallion, and causes physical changes in the cervix, uterus, and vagina to prepare the mare for conception. Estrogen peaks 1-2 days before ovulation, and decreases within 2 days following ovulation.
- *Inhibin:* Secreted by the developed follicle right before ovulation, "turns off" FSH, which is no longer needed now that the follicle is larger.
- Progesterone: Prevents conception and decreases sexual receptibility of the mare to the stallion. Progesterone is therefore lowest during the estrus phase, and increases during diestrus. It decreases about 15 days after ovulation, when the corpus luteum begns to decrease in size.
- Prostaglandin: Secreted by the endrometrium 13-15 days following ovulation, causes luteolysis and prevents the corpus luteum from secreting progesterone
- eCG (equine chorionic gonadotropin - formerly called PMS/PMSG pregnant mare serum and pregnant mare serum gonadotropin): Chorionic gonadotropins secreted if the mare conceives. First secreted by the endometrial cups around the 36th day of gestation, peaking around day 60, and decreasing after about 120 days of gestation. Also help to stimulate the growth of the fetal gonads.
- *Prolactin*: Stimulates lactation
- *Oxytocin*: Stimulates the uterus to contract

While horses in the wild mate and foal in mid to late spring, many horses domestically bred for competitive purposes, especially horse racing and various futurities, are usually born as close to January first as possible, so as to be at an advantage in size and maturity when competing against other horses in the same age group. When an early foal is desired, barn managers will put the mare "under lights" by keeping the barn lights on in the winter to simulate a longer day, thus bringing the mare into estrus sooner than she would in nature. Mares signal estrus and ovulation by urination in the presence of a stallion, raising the tail and revealing the vulva. A stallion, approaching with a high head, will usually nicker, nip and nudge the mare, as well as sniff her urine to determine her readiness for mating.

Once fertilized, the oocyte (egg) remains in the oviduct for approximately 5.5 more days, and then descends into the uterus. The initial single cell combination is already dividing and by the time of entry into the uterus, the egg might have already reached the blastocyst stage. The gestation period lasts for about eleven months, or about 342 days (normal average range 320-370 days). During the early days of pregnancy, the embryo (up to about 35 days), or fetus (after 35-40 days), is mobile, moving about in the uterus until about day 16 when "fixation" occurs. True implantation - invasion into the endometrium of any sort - does not occur until about day 35 of pregnancy with the formation of the endometrial cups, and true placentation (formation of the placenta) is not initiated until about day 40-45 and not completed until about 140 days of pregnancy. A heartbeat can be detected on day 21, and the fetus gender can be determined by day 70 of the gestation. Halfway through gestation the fetus is the size of between a rabbit and a beagle. The most dramatic fetus development occurs in the last 3 months of pregnancy when 60 per cent of fetal growth occurs.

Care of the Pregnant Mare

Domestic mares receive specific care and nutrition to ensure that they and their foals are healthy. Mares are given vaccinations against diseases such as the Rhinopneumonitis (EHV-1) virus (which can cause abortions) as well as vaccines for other conditions that may occur in a given region of the world. Mares can be used for riding or driving during most of their pregnancy, and it's healthy for them to have exercise. But only moderate exercise, especially when they become heavy in foal. Exercise in excessively high temperatures has been suggested as being detrimental to pregnancy maintenance during the embryonic period - it should however be noted that ambient temperatures encountered during the research were in the region of 100 degrees F and the same results may not be encountered in regions with lower ambient temperatures.

Mares are generally fed more than other horses and are provided supplemental vitamins and minerals because their bodies require extra nutrition to form (and later to nurse) a foal. Many feeds designed for pregnant and lactating mares also have extra fat and protein. Overfeeding the pregnant mare should also be avoided, as excess weight may contribute to difficulties foaling or fetal/foal related problems. A special foaling stall or shed that is large and clutter free provides the mare with a safe place to give birth. Most mares foal at night or early in the morning, and prefer to give birth alone when possible. Labour is rapid, often no more than 30 minutes, and from the time the feet of the foal appear to full delivery is often only about 15 to 20 minutes. Once the foal is born, the mare will lick the newborn foal to clean it and help blood circulation. In a very short time, the foal will attempt to stand and get milk from its mother. A foal should stand and nurse within the first hour of life.

To create a bond with her foal, the mare licks and nuzzles the foal, enabling her to distinguish hers from others. Some mares are aggressive when protecting their foals, and may attack other horses or unfamiliar humans that come near their newborns. After birth, a foal's navel is dipped in iodine to prevent infection, it is sometimes given an enema to help clear the meconium from its digestive tract, and the newborn is monitored to ensure that it stands and nurses without difficulty. While most horse births happen without complications, many owners have first aid supplies prepared and a veterinarian on call in case of a birthing emergency.

People who supervise foaling should also watch the mare to be sure that she passes the placenta in a timely fashion, and that it is complete with no fragments remaining in the uterus, where they could cause a serious infection. If the placenta is not removed from the stall after it is passed, a mare will often eat it, an instinct from the wild, where blood would attract predators. Foals develop rapidly, and within a few hours a wild foal can travel with the herd. In domestic breeding, the foal and dam are usually kept separate for a while, but within a few weeks are typically pastured with the other horses. A foal will begin to eat hay, grass and grain alongside the mare at about 4 weeks old; by 10-12 weeks the foal requires more nutrition than the mare's milk can supply. Foals are typically weaned at 4-8 months of age, although in the wild a foal may nurse for a year.

How Breeds Develop

Beyond the appearance and conformation of a specific type of horse, breeders aspire to improve physical performance abilities. This concept, known as matching "form to function," has led to the development of not only different breeds, but also families or bloodlines within breeds that are specialists for excelling at specific tasks. For example, the Arabian horse of the desert naturally developed speed and endurance to travel long distances and survive in a harsh environment, and domestication by humans added a trainable disposition to the animal's natural abilities. In the meantime, in northern Europe, the locally adapted heavy horse with a thick, warm coat was domesticated and put to work as a farm animal that could pull a plow or wagon.

This animal was later adapted through selective breeding to create a strong but ridable animal suitable for the heavily-armored knight in warfare. Then, centuries later, when people in Europe wanted faster horses than could be produced from local horses through simple selective breeding, they imported Arabians and other oriental horses to breed as an outcross to the heavier, local animals. This led to the development of breeds such as the Thoroughbred, a horse taller than the

Arabian and faster over the distances of a few miles required of a European race horse or light cavalry horse. Another cross between oriental and European horses produced the Andalusian, a horse developed in Spain that was powerfully built, but extremely nimble and capable of the quick bursts of speed over short distances necessary for certain types of combat as well as for tasks such as bullfighting.

Later, the people who settled the Americas needed a hardy horse that was capable of working with cattle. Thus, Arabians and Thoroughbreds were crossed on Spanish horses, both domesticated animals descended from those brought over by the Conquistadors, and feral horses such as the Mustangs, descended from the Spanish horse, but adapted by natural selection to the ecology and climate of the west. These crosses ultimately produced new breeds such as the American quarter horse and the Criollo of Argentina. In modern times, these breeds themselves have since been selectively bred to further specialize at certain tasks. One example of this is the American quarter horse. Once a general-purpose working ranch horse, different bloodlines now specialize in different events. For example, larger, heavier animals with a very steady attitude are bred to give competitors an advantage in events such as team roping, where a horse has to start and stop quickly, but also must calmly hold a full-grown steer at the end of a rope. On the other hand, for an event known as cutting, where the horse must separate a cow from a herd and prevent it from rejoining the group, the best horses are smaller, quick, alert, athletic and highly trainable. They must learn quickly, have conformation that allows quick stops and fast, low turns, and the best competitors have a certain amount of independent mental ability to anticipate and counter the movement of a cow, popularly known as "cow sense."

Another example is the Thoroughbred. While most representatives of this breed are bred for horse racing, there are also specialized bloodlines suitable as show hunters or show jumpers. The hunter must have a tall, smooth build that allows it to trot and canter smoothly and efficiently. Instead of speed, value is placed on appearance and upon giving the equestrian a comfortable ride, with natural jumping ability that shows bascule and good form. A show jumper, however, is bred less for overall form and more for power over tall fences, along with speed, scope, and agility. This favours a horse with a good galloping stride, powerful hindquarters that can change speed or direction easily, plus a good shoulder angle and length of neck. A jumper has a more powerful build than either the hunter or the racehorse. The history of horse breeding goes back millennia. Though the precise date is in dispute, humans could have domesticated the horse as far back as approximately 4500 BCE. However, evidence of planned breeding has a more blurry history.

One of the earliest people known to document the breedings of their horses were the Bedouin of the Middle East, the breeders of the Arabian horse. While it is difficult to determine how far back the Bedouin passed on pedigree information via an oral tradition, there were written pedigrees of Arabian horses by A.D. 1330. The Akhal-Teke of West-Central Asia is another breed with roots in ancient times that was also bred specifically for war and racing. The nomads of the Mongolian steppes bred horses for several thousand years as well. The types of horses bred varied with culture and with the times. The uses to which a horse was put also determined its qualities, including smooth amblers for riding, fast horses for carrying messengers, heavy horses for plowing and pulling heavy wagons, ponies for hauling cars of ore from mines, packhorses, carriage horses and many others.

Medieval Europe bred large horses specifically for war, called destriers. These horses were the ancestors of the great heavy horses of today, and their size was preferred not simply because of the weight of the armor, but also because a large horse provided more power for the knight's lance. Weighing almost twice as much as a normal riding horse, the destrier was a powerful weapon in battle. On the other hand, during this same time, lighter horses were bred in northern Africa and the Middle East by Muslim warriors, who preferred a faster, more agile horse. The lighter horse suited the raids and battles of the Bedouins, allowing them to outmaneuver rather than overpower the enemy. When Muslim warriors and European knights collided in warfare, the heavy knights were frequently outmaneuvered. The Europeans, however, soon made up for the lack of speed of their native breeds by incorporating genetic traits from captured oriental horses such as the Arabian, Barb to their stables. This cross-breeding led both to a nimbler war horse, such as today's Percheron, but also to created a type of horse known as a Courser, a predecessor to the Thoroughbred, which was used as a message horse.

During the Renaissance, horses were bred not only for war, but for haute ecole riding, derived from the most athletic movements required of a war horse, and popular among the elite nobility of the time. Breeds such as the Lipizzan were developed from Spanish-bred horses for this purpose, and also became the preferred mounts of cavalry officers, who were derived mostly from the ranks of the nobility. It was during this time that gunpowder was developed, and so the light cavalry horse, a faster and quicker war horse, was bred for a "shoot and run" tactic rather than the close hand-to-hand fighting seen in the Middle Ages. After Charles II retook the British throne in 1660, horse racing, which had been banned by Cromwell, was revived. The Thoroughbred was developed 40 years later, bred to be the ultimate racehorse, through the lines of 3 foundation Arabian stallions.

The 17th and 18th centuries saw more of a need for fine carriage horses in Europe, bringing in the dawn of the warmblood. The warmblood breeds have been exceptionally good at adapting to changing times, and from their carriage horse beginnings they easily transitioned during the 1900s into a sport horse type. Today's warmblood breeds, although still used for competitive driving, are more often seen competing in the show jumping or dressage arenas. The Thoroughbred continues to dominate the horseracing world, although its lines have been more recently used to improve warmblood breeds and to develop sport horses. The predecessor of the American Quarter Horse was developed in the 1700s, mainly for quarter racing (racing ¼ of a mile). The breed was later adapted for work in the west, and "cow sense" was particularly bred for as their use for herding cattle increased. However, because there was also a need for animals suitable for sprint racing, the modern Quarter Horse has two distinct types: the sleeker racing type and the stock horse type. The racing type most resembles the finer-boned ancestors of the first racing Quarter Horses, and the type is still used for ¼-mile races. The stock horse type, used in western events, is bred for a shorter stride, docile temperament, and cow sense. The need for horses for heavy draft and carriage work continued until the industrial revolution and the advent of the automobile and the tractor. After this time, draft and carriage horse numbers dropped significantly, though light riding horses remained popular for recreational pursuits. Draft horses today are used on a few small farms, but today are seen mainly for pulling and plowing competitions rather than farm work. Heavy harness horses are now used as an outcross with lighter breeds, such as the Thoroughbred, to produce the modern warmblood breeds popular in Olympic and sport horse disciplines. Breeding a horse is an endeavor where the owner, particularly of the mare, will usually need to invest considerable time and money.

For this reason, a horse owner needs to consider several factors, including:

- Does the proposed breeding animal have valuable genetic qualities to pass on?
- Is the proposed breeding animal in good physical health, fertile, and able to withstand the vigours of reproduction?
- For what purpose will the foal be used?
- Is there a market for the foal in the event that the owner does not wish to keep the foal for its entire life?
- What is the anticipated economic benefit, if any, to the owner of the ensuing foal?
- What is the anticipated economic benefit, if any, to the owner(s) of the sire and dam or the foal?
- Does the owner of the mare have the expertise to properly manage the mare through gestation and parturition?

- Does the owner of the potential foal have the expertise to properly manage and train a young animal once it is born?

There are value judgements involved in considering whether an animal is suitable breeding stock, hotly debated by breeders. Additional personal beliefs may come into play when considering a suitable level of care for the mare and ensuing foal, the potential market or use for the foal, and other tangible and intangible benefits to the owner. If the breeding endeavor is intended to make a profit, there are additional market factors to consider, which may vary considerably from year to year, from breed to breed, and by region of the world. In many cases, the low end of the market is saturated with horses, and the law of supply and demand thus allows little or no profit to be made from breeding unregistered animals or animals of poor quality, even if registered.

The minimum cost of breeding for a mare owner includes the stud fee, and the cost of proper nutrition, management and veterinary care of the mare throughout gestation, parturition, and care of both mare and foal up to the time of weaning. Veterinary expenses may be higher if specialized reproductive technologies are used or health complications occur. Making a profit in horse breeding is often difficult. While some owners of only a few horses may keep a foal for purely personal enjoyment, many individuals breed horses in hopes of making some money in the process.

A general rule of thumb is that a foal intended for sale should be worth three times the cost of the stud fee if it were sold at the moment of birth. From birth forward, the costs of care and training are added to the value of the foal, with a sale price going up accordingly. If the foal wins awards in some form of competition, that may also enhance the price.On the other hand, without careful thought, foals bred without a potential market for them may wind up being sold at a loss, and in a worst-case scenario, sold for "salvage" value — a euphemism for sale to slaughter as horsemeat.

Therefore, a mare owner must consider their reasons for breeding, asking hard questions of themselves as to whether their motivations are based on either emotion or profit and how realistic those motivations may be.The stallion should be chosen to complement the mare, with the goal of producing a foal that has the best qualities of both animals, yet avoids having the weaker qualities of either parent. Generally, the stallion should have proven himself in the discipline or sport the mare owner wishes for the "career" of the ensuing foal. Mares should also have a competition record showing that they also have suitable traits, though this does not happen as often. A purebred horse is usually worth more than a horse of mixed breeding, though this matters more in some disciplines than others.

The breed of the horse is sometimes secondary when breeding for a sport horse, but some disciplines may prefer a certain breed or a specific phenotype of horse. Sometimes, purebred bloodlines are an absolute requirement: For example most Racehorses in the United States must be recorded with a breed registry in order to race. Bloodlines are often considered, as some bloodlines are known to cross well with others. If the parents have not yet proven themselves by competition or by producing quality offspring, the bloodlines of the horse are often a good indicator of quality and possible strengths and weaknesses.

Some bloodlines are known not only for their athletic ability, but could also carry a conformational or genetic defect, poor temperament, or for a medical problem. Some bloodlines are also fashionable or otherwise marketable, which is an important consideration should the mare owner wish to sell the foal. Horse breeders also consider conformation, size and temperament. All of these traits are heritable, and will determine if the foal will be a success in its chosen discipline. The offspring, or "get," of a stallion are often excellent indicators of his ability to pass on his characteristics, and the particular traits he actually passes on. Some stallions are fantastic performers but never produce offspring of comparable quality. Others sire fillies of great abilities but not colts. At times, a horse of mediocre ability sires foals of outstanding quality. Mare owners also look into the question of if the stallion is fertile and has successfully "settled" (i.e. impregnated) mares. A stallion may not be able to breed naturally, or old age may decrease his performance. Mare care boarding fees and semen collection fees can be a major cost. Breeding a horse can be an expensive endeavor, whether breeding a backyard competition horse or the next Olympic medalist.

Costs may include:

- The stud and booking fee
- Fees for collecting, handling, and transporting semen (if AI is used and semen is shipped)
- *Mare exams*: to determine if she is healthy enough to breed, to determine when she ovulates, and (if AI is used) to inseminate her
- Mare transport, care, and board if the mare is bred live cover at the stallion's residence
- Veterinary bills to keep the pregnant mare healthy while in foal
- Possible veterinary bills during pregnancy or foaling should something go wrong
- Veterinary bills for the foal for its first exam a few days following foaling

Stud fees are determined by the quality of the stallion, his performance record, the performance record of his get (offspring), as well as the sport and general market that the animal is standing for. The highest stud fees are generally for racing Thoroughbreds, which may charge from two to three thousand dollars for a breeding to a new or unproven stallion, to several hundred thousand dollars for a breeding to a stakes winner. Sport horse stallions generally range from $1000 to $3000, although the top stallions may reach $4000 for one breeding.

The lowest stud fees may only be $100-$200, but there are trade-offs: the horse will probably be unproven, and probably much less athletic than a horse with a stud fee that is a bit higher. As a stallion's career, either performance or breeding, improves, his stud fee tends to increase in proportion. If one or two offspring are especially successful, winning several stakes races or an Olympic medal, the stud fee will generally greatly increase. Younger, unproven stallions will generally have a lower stud fee earlier on in their careers.

To help decrease the risk of financial loss should the mare die or abort the foal while pregnant, many studs have a live foal guarantee (LFG), allowing the owner to have a free breeding to their stallion the next year. However, this is not offered for every breeding.

6

Mating Behaviour in Animals

There, one lineage is currently being lost because males of that lineage are more timid and less able to defend females than those of the more successful line. Knowledge of patterns of mating and the cues that individuals use to select mates can also help identify the potential for hybridization between species; because one of the goals of conservation is species integrity and persistence, hybridization is normally viewed with alarm. Greene, Umbanhowar, Mangel, and Caro show that in infanticidal species where males kill offspring fathered by other males, population growth rates are reduced, and it may be difficult for such species to withstand exploitation. In theory, Ne will be lowered in species in which infanticide or mate guarding is prevalent.

Different fighting tactics are usually associated with different mortality schedules; thus the frequency with which males display different behavioural and morphological tactics will affect population growth rate and Ne. Rubenstein draws attention to the fact that those Atlantic salmon *Salmo salar* migrating to the sea are more likely to be caught by fishermen than conspecifics that remain behind and mature in their natal streams and that this differential offtake selects for philopatric phenotypes.

In a number of reef-dwelling fishes, intersexual competition for mating opportunities has led to the evolution of protandry, in which individuals change from male to female, or protygyny, in which they change from female to male. Sex change may be precipitated by reaching a certain age or size or by social circumstances when it pays individuals to become the currently rarer sex. Vincent and Sadovy argue that exploitation of larger fish for the tropical fish trade may remove males of certain species, for example, from a population. Thus the speed with which females are able to change to being male will be crucial in determining a population's response to exploitation.

MATE CHOICE

Among females, mate choice is increasingly seen as a means to enhance offspring fitness through choice of males with heritable, condition-enhancing traits such as resistance to parasites. In the wild, loss of males from small populations may theoretically limit the range of males from which to choose to the extent of slowing the rate of population increase. In captive populations, some degree of female choice may be advisable because, as managers, we are poor at picking out healthy stock, at identifying traits that enhance fitness, and we have little to no knowledge of trait heritability. Grahn, Langefors, and von Schantz argue that failing to allow for mate choice in captive salmon stocks has resulted in disease-induced mortality.

Mate choice has possible implications for captive breeding of endangered birds, too. Zebra finches *Poephila guttata* show preferences for mating with individuals wearing particular colour bands and will adjust the amount of parental investment and sex ratio of their offspring according to the colour of their partner's leg bands. This may upset population projections of *in situ* breeding Programmes. Recently, considerable attention has focused on the relationship between mate choice and the degree of symmetry in male phenotypic traits. This is because symmetrical traits are thought to reflect developmental stability and hence the ability of the genotype to control development under a range of environmental conditions. There is evidence that symmetrical ornaments correlate with fitness and that females choose males on the basis of ornament symmetry. Although behavioural ecologists are interested in trait asymmetry as a marker of genetic quality, asymmetry also reflects the extent of environmental insult during development and, measured repeatedly, can provide an index of a population's response to environmental change over time. Furthermore, asymmetric leg bands placed on captive male birds may modify female choice and could result in lower reproductive output in endangered species.

MATING SYSTEMS

The distribution of resources crucial to females, and hence the economic defendability of females by males interact with patterns of parental care to produce the mating system. Because Ne is lowered by increasing skew in the ratio of breeding females to breeding males, whether a population is polygynous or monogamous, or more specifically, the extent of extrapair paternity, therefore has a strong bearing on loss of genetic diversity. In addition, Ne is lowered by increasing variance in family size, which is marked in reproductively suppressed species. Creel systematically explores aspects of the mating system that affect Ne using demographic data on wild carnivores. He reports that unequal reproductive success causes the greatest reduction in Ne, with skewed sex ratios and variation in population size being less important.

Differences in mating system will also affect a population's ability to sustain commercial exploitation or harvesting. As an example, population growth rates of monogamous species are likely to be less resilient to removal of individuals than are those of polygynous species because in monogamous species there are fewer males available for remating and parental care.

PARENTAL CARE AND SEX ALLOCATION

Theory underlying the evolution of parental care, including intraspecific variation in care in relation to costs and benefits, parent-offspring conflict, and differential investment in sons and daughters, is well understood.The sex that cares for offspring and the period of time over which care occurs will affect a species' ability to withstand offtake of a given magnitude and selectivity. Vincent and Sadovy suggest that loss of parental males could reduce the reproductive output of an exploited population in taxa such as seahorses in which male care limits female reproductive rate. Adaptive alterations in offspring production can affect conservation agendas in at least two ways. First, in populations that are selectively exploited, parents may bias offspring production in favour of the rarer sex. For example, Creel and Creel have shown that in lions birth sex ratios are male biased in populations hunted for male trophies, and this has ramifications for sustainable offtake. Rubenstein shows how failure to account for sex ratio adjustment in onagers *Equus hemionus* jeopardized a reintroduction Programme. Second, bird recovery Programmes make use of a female's disposition to lay replacement clutches if her first brood is removed, and they also capitalize on hatching asynchrony to cross-foster the first hatchling to a conspecific or heterospecific.

Both management strategies can improve reproductive output of endangered species. In addition, knowledge of siblicide in birds is useful in preventing reproductive loss when trying to breed endangered birds. In theory, one might expect the rate and distance over which animals disperse and its associated costs to have greater ramifications for conservation strategies than any other behavioural ecological variable. Rates of dispersal between habitat patches affect population persistence gene flow between subgroups influences heterozygosity and genetic drift, and the length and direction dispersers travel and their habitat preferences during movement speak to reserve design and connectivity.

Although dispersal is often studied in animal and plant populations high-quality data are, unfortunately, difficult to obtain; for example, maximal dispersal distances are usually underestimated in behavioural studies. Outside of PVAs, there have been few attempts to use empirical data on dispersal in conservation theory or practice. To remedy this Van Vuren outlines the way in which dispersal distance, dispersal direction, and survival of dispersers might influence design of reserves and managing areas between them. As populations become increasingly isolated, the threat of inbreeding increases. Knowledge of the extent of philopatry and strength of inbreeding avoidance mechanisms across taxa can identify which species are most likely to be subject to inbreeding depression. This would then prioritize which species or populations require translocations and on which species noninvasive genetic screening might be carried out, as advocated by Waldman and Tocher. In addition, the extent to which different species tolerate inbreeding in the wild informs captive Programmes about the relative importance of avoiding pairing relatives in zoo breeding Programmes.

COOPERATIVE BREEDING

Many examples of cooperative breeding in birds and mammals are a consequence of habitat saturation, a situation increasingly likely to occur as habitats become isolated and smaller in area through fragmentation. Populations of cooperative breeders are likely to have lower population growth rates than populations where all individuals pair up.

For instance, Seychelles warblers *Acrocephalus sechellensis* transferred from densely populated Cousin Island to empty Aride Island exhibited faster rates of reproduction than their founding population. On the other hand, the presence of helpers or floating nonresident individuals can act as a buffer to small decreases in population size, but the dynamics of these conflicting effects are unexplored. Knowledge of the circumstances under which cooperative breeding will appear in different species may therefore inform intervention strategies, while reproductive benefits that parents receive from offspring can help zoo breeding Programmes set up appropriate social groups. Behavioural ecology focuses on the adaptive significance of variation in individual behaviour, morphology, and physiology which has forced scientists to learn to recognize individual animals.

Consequently, all the individuals within a subpopulation are often known to researchers fortuitously providing an accurate estimate of local population size. Individually recognizing all members of a population is a far more accurate method of estimating population size than using classical ecological techniques such as mark-recapture methods or stratified sampling. In addition, long-term field studies have unusually accurate figures of changes in population size in turn, this may influence effective population size. McGregor and Peake review methods used to recognize animals individually and additionally show that populations can be monitored through individual call recognition rather than observing animals themselves. Knowledge of the number of lineages, or breeding groups, for a given Ne will alter estimates of the time frame over which allelic diversity is maintained. Allozymes, DNA fingerprinting, mitochondrial DNA (mtDNA), minisatellite, and microsatellite techniques shed light on population differentiation and genetic variability.

These methods have identified subpopulations of humpback whales *Megaptera novaeangliae* migration routes and extent of philopatry in marine turtles, and the degree of hybridization between canid species information almost impossible to obtain from observation alone. In addition, analyses of mtDNA sequences determine phylogenetic relationships between species, which help identify evolutionarily significant units (ESUs) for conservation. For instance, genetic analyses of orangutan *Pongo pygmaeus* populations indicate sufficient divergence between Sumatran and Bornean populations to put a halt to breeding individuals from the two regions.

Although within-and between-species molecular comparisons are normally outside the purview of behavioural ecology, analyses that attempt to uncover associations between behavioural and morphological variables without the confounding effects of shared ancestry nonetheless require phylogenetic trees derived using molecular techniques. Because anthropogenic factors are the greatest challenge to biodiversity, many conservation biologists would argue that conservation solutions will eventually emerge from understanding human behaviour.

Patterns of human exploitation are better understood now than they were 15 years ago. The myth of the "noble savage" living in harmony with the environment has been exploded. In a detailed exploration of this issue, Alvard uses empirical data to show that Piro hunters in Peru pursue an optimal foraging strategy rather than any conservation agenda. As a second illustration, the tragedy of the commons, which describes individuals overexploiting a common property, is better understood in terms of a problem of unrestricted access rather than a problem of ownership.

This focuses attention on the best ways to police resources rather than on who owns them. These findings are informative in that they show which conservation strategies are likely to succeed or fail. In spite of these advances, the rapid growth of evolutionary psychology, a subdiscipline in which contemporary human behaviour is interpreted in the light of past selection pressures, has so far had little to say about patterns of current exploitation.

Similarly, economics does not consider the underlying biological motivation in decision making. Nevertheless, it seems likely that we will be able to understand human resource use better by paying closer attention to evolved behavioural dispositions. Specifically, they draw attention to differences in attitudes towards conservation shown by different segments of society as predicted by evolutionary theory. Certainly, the two pivotal problems in conservation biology, human population growth and resource accrual, are central to human behavioural ecology's focus on the social and ecological factors that affect reproductive success in different cultures. Demographic records and information on dispersal are obviously of great importance, while population growth rate, a population's response to exploitation, and Ne are influenced by many factors in behavioural ecology.

There are other bridges between these disciplines that have not been included, however, and many others that doubtless will appear as our knowledge of behavioural adaptations improves. Moreover, we think it fair to say that any behavioural ecologist with a conscience must be interested in the disappearance of biological diversity and, with a little thought, should be able to recognize the conservation significance of his or her research and make some contribution to solving the crisis at hand.

IMPORTANCE OF CAPTIVE BREEDING

The importance of captive breeding as a viable tool for conservation of rare and endangered species has been emphasized by zoos, aquariums, and other captive facilities since the early 1970s. Zoos have repeatedly been portrayed as future "arks" for endangered species. In concordance with this perspective, the International Union for the Conservation of Nature (IUCN) has adopted a policy endorsing captive breeding as an important supportive intervention to avoid the loss of many species. The policy calls for the establishment of captive-breeding Programmes and self-sustaining captive populations (*i.e.*, populations that are viable in captivity without further genetic influx from the wild) before a species is reduced to critically low numbers in the wild. Captive-breeding Programmes are to be coordinated internationally according to genetic and demographic principles. The ultimate goal of captive breeding Programmes is the maintenance and/ or reestablishment of viable populations in the wild.

Although there is general agreement that as more species become extinct in the wild, more will persist only in captivity the effectiveness and relative importance of captive-breeding Programmes for overall conservation efforts remains under debate. Today, there have been few success stories of species rescue and reintroduction. So far, 25 species have been preserved in captivity ollowing extinction in the wild. Among these are eight mammal species, six of which have been reintroduced into areas of their former range: European bison *Bison bonasus;* Przewalski *horse Equus przewalskii;* Pere David deer *Elaphurus davidianus;* Arabian oryx *Oryx leucoryx;* red wolf *Canis lupus;* and black-footed ferret *Mustela nigripes.* In addition, wild populations of a number of threatened mammal species have been reinforced by captive breeding. Despite considerable financial costs and animal losses, some of these Programmes are conservatively regarded as successful.

Magin et al. recently estimated that specimens of approximately 34 per cent of all threatened mammal species are held in zoos worldwide, and about 15 per cent of the zoos' 200,000 "mammal spaces" are devoted to threatened taxa.

However, only 90 of the 174 zooheld threatened species are kept at levels that could allow establishment of self-sustaining captive populations that would be viable in the absence of future genetic interchange. Similarly, a perusal of recent reports on North American Species Survival Plans (SSP) of mammals showed that only 13 of the 56 species and subspecies managed under North American SSP Programmes had self-sustaining or stable captive populations. Two commonly encountered difficulties in attaining viable captive populations are unreliable or no reproductive success and high rates of infant mortality. The question then is, how can we solve existing breeding problems and achieve sustainability of captive populations to turn captive breeding into a useful and reliable conservation tool where needed? Over the past 10 years, the captive-breeding community has evolved into a powerful and influential component of the world's conservation scene.

Many of today's zoos and captive breeding facilities changed from simple menageries, through a period of living museums with more naturalistic exhibits, to today's modern conservation centers with a vested interest in preserving *in situ* (*i.e.*, in the wild) biodiversity. The recent publication of a World Zoo Conservation Strategy by the World Zoo Organization (IUZDG) and the Conservation Breeding Specialist Group (CBSG) of the Species Survival Commission of the IUCN emphasizes broad goals for zoo conservation and the integration of in situ and ex situ (*i.e.*, in captivity) conservation efforts. Zoo contributions to conservation efforts may be grouped into four major areas.

First, zoos can educate and influence a largely urban public to appreciate the value and the importance of species and ecosystem conservation. This may represent their most powerful contribution to conservation efforts. An estimated 100 million people visit North American zoos every year, and approximately 600 million people, about 10 per cent of the world's population, are thought to visit zoos worldwide per year. Second, zoos can contribute substantially to *in situ* conservation efforts through fundraising and sponsoring field projects. The use of endangered mammals as flagship species for *in situ* and *ex situ* protection efforts may be most effective to secure monetary, political, and public support for conservation efforts. Some organizations such as the Frankfurt Zoological Society, the New York Wildlife Conservation Society, the National Zoological Park of the Smithsonian Institution, and the San Diego Zoological Society, for example, have been involved in supporting field research and *in situ* conservation efforts for many years. Now smaller zoos are also discovering the benefits of such projects. Third, basic research carried out in captive facilities can contribute substantially to our understanding and knowledge of biological processes. Rigorous research on captive populations of endangered species may also provide us with improved knowledge for management of wild populations and help us to develop new technologies useful for conservation efforts.

The fourth, and over the past decade, most advertised, contribution is the envisioned role of zoos as "arks" for species survival by providing self-sustaining captive populations of endangered species and subspecies for future reintroduction into the wild. The frequently stated goal for captive populations is to preserve 90-95 per cent of heterozygosity for 100-200 years. The rationale for long-term captive propagation is based on the assumption that human population growth may level off in about 200 years, and restoration of habitat and species may then become possible. Critics, however, have pointed out that zoos do not have sufficient space to accommodate viable captive populations for all threatened species, and the costs for captive-breeding Programmes can be staggering, up to a half-million dollars per year per species. More recently, therefore, the so-called Noah's ark paradigm of captive breeding has been amended to provide a more realistic conservation goal which attempts to consider spatial and financial limitations in captive facilities.

In this new endeavor, captive-breeding Programmes are envisioned as participating in an interactive management strategy largely focused on *in situ* conservation efforts with coordination on a global scale. Today's zoo conservation efforts have become a systematic and cooperative venture based on scientific principles largely derived from the areas of demography and population genetics. A complex bureaucracy with a dizzying array of acronyms, representing various associations, committees, groups, and conservation tools, has developed as part of this worldwide effort.

Systematic census and inventory Programmes, such as the Animal Records Keeping System and the International Species Information System, are used to compile information on captive populations of all species from the majority of the world's large zoos. To coordinate these efforts and to integrate collected information the Conservation Breeding Specialist Group, a multidisciplinary specialist group, was organized by the IUCN Species Survival Commission. The CBSG advises, monitors, and catalyzes captive propagation on a global basis and provides a link between *in situ* and *ex situ* conservation efforts worldwide. However, genetic and demographic issues have so far predominated captive breeding management and have provided the foundation for captivebreeding plans.

Now the need for a broader approach to captive management has been recognized and research efforts incorporating behaviour, nutrition, disease, physiology, genetics, population biology, and various interdisciplinary studies are seen as necessary to facilitate responsible and successful captive propagation and conservation. An examination of North American SSP reports and other publications relevant to the zoo community revealed that behavioural problems are frequently cited as major impediments to Also, much evidence suggests that behavioural research has been instrumental in past breeding and husbandry improvements. However, little information is available on the social and behavioural requirements and general biology of many endangered species when breeding Programmes are initiated. Solutions to captive breeding problems are often found by trial and error and by exchange of anecdotal and descriptive information among facilities rather than by systematic investigation. Behavioural studies that address the conservation needs of captive-breeding

Programmes using systematic hypothesis testing are rare. As a consequence, animal behavioural research is seldom incorporated in *ex situ* conservation planning, and no general theory exists on which aspects of behaviour are likely to be the most critical for conservation efforts. Integrating systematic behavioural and husbandry research into conservation planning has been highlighted as crucial for *ex situ* conservation.

ZOO CONSERVATION

Behavioural studies have predominated in zoos for decades. Animal behaviour studies have contributed significantly to the improvement of management and reproduction of zoo animals and have been regarded as a powerful tool for captive management. However, many behavioural research papers, although important in their findings, are the result of casual observation or purely descriptive case studies. They frequently arise from an effort to solve a particular crisis encountered with captive-held individuals or species. Solutions are applied over time until breeding success is obtained, and biological understanding of the obtained solution may or may not be achieved during this process. The scientific method using systematic hypotheses testing has rarely been employed in such studies. Recent technological advances in wildlife immobilization and anesthesiology, biotechnology, genetics, and reproductive physiology have received increasing attention in the search for solutions to breeding problems.

The development of techniques for assisted reproduction (*e.g.*, artificial insemination, *in vitro* fertilization, and embryo transfer) has contributed much valuable information to our overall understanding of the reproductive biology and physiology of many species. Furthermore, in some cases the use of assisted reproduction may provide the only feasible way to save a species or subspecies from certain extinction. Theoretically, techniques for assisted reproduction can offer a way to increase species capacity and to facilitate the movement of animals without the hazards of disease, behavioural stress, and social problems. Similarly, importing genetic material from wild into captive populations can be accomplished without further depletion of wild specimens. However, so far these technologies have only been worked out for less than three dozen wild species of mammals and birds, and broader application may not be feasible due to limited resources. Furthermore, for species in which such technologies are available, concurrent behavioural studies are urgently needed to examine possible drawbacks of the use of assisted reproduction. Shepardson appropriately cautioned that there is more to preserving species than looking after their genes. The intended goal of captive-breeding Programmes is to preserve animals as representatives of their wild ancestors for posterity, research, education, and for potential release back to the wild. Thus, preservation of natural behaviours should be a key factor for all these endeavors.

Failure to reproduce naturally and raise young in the captive environment will most likely result in a loss of many forms and patterns of natural behaviour. Many mammalian species show a large repertoire of complex learned and culturally transmitted behaviours (*e.g.*, sexual and parental competence, social interactions, some aspects of foraging behaviours). Such behaviours may be lost permanently if no provisions are made to support their development and display in captivity. Indeed, the diversity of culturally transmitted behaviours has the potential to be lost much faster than genetic diversity.

It is therefore of utmost importance that captive animals develop behavioural competence in all areas important for living in the wild if the ultimate goal is reintroduction. Provision of more naturalistic environments (*i.e.*, simulating a species' social and physical environment found in the wild) may help to minimize the effects of domestication or selection for a specific captive environment, and successful unassisted captive propagation of a species may serve as an indicator of our intuitive or scientific understanding of a particular species' requirements. In short, systematic research efforts are urgently needed to examine various aspects of behaviour pertinent to conservation efforts for captive and wild populations of rare and endangered species. Below, several areas of behavioural research that are particularly relevant for captive-breeding success in mammals are identified. Examples of successes and failures of captive propagation that can be attributed to behavioural factors and that illustrate the importance of behavioural research are presented in each section.

The chosen examples are based on an extensive review of the zoo literature with particular emphasis on the *Journal of Zoo Biology* and the *International Zoo Yearbook*. Hediger first noted the importance of territoriality and functional space for captive management of wild animals. Knowledge of species-specific movement patterns and territory requirements is essential for maintaining close-to-natural social and reproductive interactions in mammals. In addition, perceived space, including olfactory and auditory components, should be considered together with specific niche requirements for successful maintenance and breeding of captive mammals. Providing appropriate space objects (*e.g.*, climbing structures, hiding places) has been shown to have substantial behavioural effects by reducing aggression and stress in several species.

For example, in a pair of giant pandas *Ailuropoda melanoleuca*, a significant decrease in frequency of agonistic vocalizations occurred after addition of a complex series of platforms, swinging tires, and other mobile features to the enclosure. In leopard cats *Felis bengalensis*, reduced exploratory behaviour was an indicator of chronically elevated adrenocortical activity (indicating stress), and hiding behaviour increased with elevated adrenocortical output. Consequently, provision of appropriate hiding places stimulated exploratory behaviour and reduced pacing and urinary cortisol levels.

These results indicate that structural enrichment of enclosures may increase behavioural opportunities for coping with aversive stimuli. Groups of rodents and nonhuman primates in captivity exhibited higher frequencies of aggressive behaviour than less crowded or free-living conspecifics. Blake and Gillett found that an increase in cage size resulted in larger litter sizes and increased overall breeding success in Asian chipmunks Tamias sibiricus. In other rodents, cannibalism and delayed implantation have been observed as a result of crowding.

Even when cage size is appropriate, sounds and odors from adjacent enclosures may lead to a perceived crowding effect or territory invasion reported that in an off-exhibit breeding facility for golden lion tamarins *Leontopithecus rosalia* where an air-handling system inadvertently also moved odors and sounds from enclosure to enclosure, neighbouring family groups frequently seemed agitated and vocalized in a distressed manner. Among primates an inappropriate captive environment is frequently associated with stereotyped movement, self-aggression, higher levels of aggression between cage mates, inactivity, and infertility. Indeed, numerous studies on a variety of mammal species have documented an increase in natural behaviour patterns and a decline is stereotypic movements after providing various enrichment devices and increasing the complexity of the captive environment.

In some cases reproduction may be directly and positively affected by such changes. For example, several previously nonreproductive gorillas (*Gorilla*) became reproductively active after transfer to an enriched enclosure. For some species certain enclosure features may be necessary simply to increase survivorship in captivity. For example, in several ungulate species fatalities have been reported due to inadequate enclosures or enclosure size. Death resulted from collision with fences or other enclosure structures. In pronghorn *Antilocapra americana* the importance of distant-visibility has been emphasized, and recommendations for fencing and shelter constructions have been made to accommodate natural behaviours and reduce fatalities. Furthermore, species-specific enclosure design needs to consider appropriate hiding places, denning sites, substrate, and other structural features. For example, the giant panda spends little time auto-grooming but requires soil or grass and water to allow for regular rubbing and bathing to maintain a healthy fur condition.

INTERSPECIFIC INTERACTIONS

Some species are maintained in mixed-species exhibits. This practice is particularly common in ungulates. However, excessive aggression and even fatalities have sometimes occurred as a result. Zebras, for example, have been observed to kill wildebeest and giraffe calves at some facilities. In a multi-institutional study on interspecific aggression in 10 ungulate species, a correlation between aggressive behaviour and taxonomic relatedness was found.

Interestingly, aggression was found to be highest between distantly related species. A possible explanation offered for this phenomenon was the inability of distantly related species to recognize each others' threat displays and thus avoid further aggressive interaction. However, there are encouraging accounts of various attempts to establish mixed-species exhibits, and some have reported an increase in the behavioural repertoire shown by species under mixed-housing conditions.

For small to medium-sized predators, such as cheetahs *Acinonyx jubatus* and a number of small felid species, it has been reported that housing next to large-sized predators and natural competitors, such as lions *Panthera leo* and leopards *Panthera pardus*, may decrease reproductive success. Although recently there has been increased awareness that some species may not thrive well in captivity if in constant, close proximity to their natural predators there have been few systematic studies to investigate this contention. One study indicates that ungulates may change their behaviour as a result of constant visual presence of predators. However, the authors concluded that because the changes only occurred in rare behaviours, juxtaposing predators and their prey in exhibits did not result in serious behavioural changes.

FORAGING BEHAVIOUR

Wild mammals usually spend a significant amount of time foraging. For instance, wild orangutans *Pongo pygmaeus* spend about 46 per cent of their time feeding, wild gorillas about 45 per cent and chimpanzees *Pan troglodytes* about 74 per cent.

During foraging wild animals face decisions about where, when, and how long to forage, and which items to choose. However, most captive mammals are deprived of such experiences and are frequently provided with uniform diets presented at the same time and place every day.

At one facility foraging time in captive gorillas was reported to have increased from 11 per cent to 27 per cent after the provision of browse; another study on captive gorillas reported about 8 per cent of time spent on feeding and foraging. It is likely that sev erely reduced levels of foraging behaviour are prevalent for most captive mammals when compared with their wild counterparts, and variation of food presentation over space and time may allow increased foraging activity in captive specimens. Foraging may be associated with developmental, parental, and sexual behaviour for many species. Provision of more varied diets and the possibility to exhibit natural foraging behaviour may therefore be important for breeding success. Breeding success may also be influenced indirectly by enhancing the general well-being of the animal or by directly stimulating sexual activity.

For example, exhibition of more complex feeding behaviours decreases stereotypic behaviours and increases general activity and well-being in different species such as macaques, orangutan several other primate species, and leopards. In the cheetah the stimulation provided by catching and killing live prey appears to enhance breeding success in some instances. Hediger noted that feeding wild animals in captivity should meet nutritional requirements but also needs to fulfill psychological and behavioural requirements. Detailed knowledge about natural foraging behaviour and foraging patterns is necessary to provide appropriate variety and appropriate temporal and spatial presentation of food items.

Aggression in a group of stump-tailed macaques *Macaca arctoides* was reduced by presenting food scattered on to a wood-chip litter substrate. Similar results were obtained for pig-tailed macaques *Macaca nemestris* and for chimpanzees. In the light of preserving natural behaviours for possible reintroduction, the importance of learning species-specific foraging patterns may be crucial for success. For example, adult dietary preferences of black-footed ferrets were found to be established at an early developmental stage during the third month of life.

FORMATION AND DISPERSAL OF GROUP

The importance of providing an appropriate social environment for successful captive propagation has frequently been emphasized. Nevertheless, many species are still held in arrangements that consist of forced social living of at least one adult of each gender being permanently housed together. Species that require different social conditions may show little to no reproductive success or will be unable to rear viable offspring as a consequence.

Our understanding of the social interactions of endangered species is frequently based on anecdotal information or few rigorous studies. For example, most felids are simply regarded as asocial, yet data from field studies only exist for a small number of felid species. In one case, extensive field studies revealed an unexpected pattern of sociality given the species' solitary nature: in the cheetah, males formed long-term stable bonds termed "male coalitions". Thus, patterns of sociality even in largely solitary species may be more intricate than assumed from minimal observations or extrapolation of findings in related species. Furthermore, extrapolation from closely related species may frequently prove inadequate.

For example, three zebra species, Burchell *zebra Equus burchelli,* Grevy *zebra Equus grevyi,* and mountain zebra *Equus zebra,* were found to behave quite differently from each other with regard to herd formation. False assumptions about the social nature of an endangered species, based on extrapolations and scant field data, are not uncommon, and detection of such errors is critical for long-term captive-breeding success. For example, the aye-aye *Daubentonia madagascariensis* was originally described as solitary, but later findings indicated a more social nature. Matschie's tree kangaroos *Dendrolagus matschiei* were traditionally housed in pairs or small groups, but behavioural studies on captive and wild tree kangaroos suggest a largely solitary existence. The brown hyena *Hyaena brunnea,* although belonging to its own family (*Hyaenidae*), had mostly been kept in pairs, a social structure successful for many canid species; however, field studies showed that this species is far more social than had been assumed. Elephant shrews *Elephantulus rufescens* were often kept in groups but never reproduced in this situation. Field research indicated that most elephant shrew species might be monogamous. Once housing arrangements in captivity had been changed and elephant shrews were kept in pairs rather than in groups, successful reproduction occurred. Similarly, red pandas *Ailurus fulgens* were found to be unable to rear offspring when two or more females were housed together. Among marmoset and tamarin species, founding pairs were often allowed to develop into large family groups.

The offspring, however, would never reproduce in these family settings. Periodic removal of adolescent young and their establishment as new pairs resulted in release of reproductive suppression and was therefore crucial for a successful breeding Programme. Various types of reproductive suppression have been documented for a number of species, where either offspring remained within family groupings past the time of sexual maturity — *e.g.*, bush dog *Speothos venaticus,* cottontop tamarin *Saguinus oedipus,* naked mole rat *Heterocephalus glaber* or where individuals with low dominance status were forced to remain in inappropriate social settings — *e.g.*, black-howler monkey *Aloutta caraya* cotton-top tamarin Japanese macaques *Macaca fuscata,* marmosets and tamarins.

This may be part of a reproductive strategy for some species, where alloparental "helpers" are necessary to rear offspring successfully in the wild. However, for other species it may be a reflection of inappropriate social conditions in captivity. Therefore, detailed knowledge of natural patterns of dispersal and grouping is necessary to determine species-specific management decisions. Inappropriate social composition has also been observed to lead to abortion or unsuccessful reproduction. Furthermore, introductions of unfamiliar individuals into established groups without knowledge of dispersal patterns in the wild may lead to serious injuries, unexpected fatalities, or death of the newcomer. For some ungulates, the establishment of stable female hierarchies has proven crucial for reduced aggression and successful reproduction. Furthermore, there is evidence that a minimum group size may be necessary for successful reproduction in some species. Such social facilitation of reproduction, also termed the Allee effect has been documented for several bird species. However, it may also occur in mammals. For example, in black lemurs *Eulemur macaco* and mongoose lemurs *Eulemur mongoz,* reproductive success may be enhanced by the presence of additional conspecific pairs. In Przewalski's horse, white rhinoceroses *Ceratotherium simum,* and vicunas *Vicugna vicugna* larger groups have been found to dissipate aggression of males and avert fatalities in females, and they may also act as a stimulus necessary for breeding.

REPRODUCTIVE BEHAVIOUR

Consideration of species' mating systems is one of the most basic requirements of captive propagation. Historically, however, it has been common practice in zoos to house only a single pair, male and female, of each species. Mating systems in mammals are highly diverse; 11 different mating systems have been described in one classification and only about 3 per cent of mammals have been classified as monogamous. Thus, monogamous pairings were only successful for a small number of species and individuals. In many species the natural mating system may not be obvious from the social composition observed in the wild. For example, in packs of wolves and African wild dogs, only one pair will form a monogamous unit and reproduce; however, to a casual observer, polygygny, polyandry, or promiscuity may be inferred based on social composition alone.

Even when the general mating system is known (*e.g.*, giant panda, cheetah), our understanding of the underlying behavioural-physiological mechanisms of mate location, mate choice, timing and induction of sexual arousal, and courtship behaviours may be insufficient to create the appropriate environment for successful reproduction. This is particularly apparent in asocial species, where mating represents an unusual event of "sociality" during which the tendencies to avoid conspecifics are overcome for a short time. Mating in these species may involve a substantial risk, and aggression is usually part of the courtship ritual. There are numerous accounts of captive animals where mates have killed or severely injured each other. Behavioural studies have offered successful solutions in some cases. For example, Thompson's study of captive kangaroo rats suggested that individuals destined for breeding should be kept within sensory contact of the opposite sex to allow for socialization and familiarization.

Long-term familiarity appears to mitigate aggression among males and females. Similarly, in clouded leopards and leopards, one strategy of achieving compatible breeding pairs is to place sexually immature individuals together for bonding and familiarization. However, long-term familiarity of breeding partners can also be detrimental to breeding success, particularly in species where avoidance of mating with closely familiar individuals may be part of a strategy to avoid inbreeding or to facilitate dispersal. Introductions of an unfamiliar estrous female can lead to sudden sexual interest in previously sexually unresponsive males. For the cheetah, an important provision for repeated successful breeding is to isolate females and only occasionally bring males and females together for breeding introductions.

More generally, the presentation of novel stimuli such as live prey or an altered environment may also stimulate reproductive activity in some species (*e.g.*, cheetah, red panda, white rhinoceros). Courtship chases or courtship-associated fighting may also be a necessary part of reproduction (*e.g.*, cheetah: black rhinoceros). For several gregarious ungulate species (*e.g.*, white rhinoceros, Asiatic wild ass *Equus hemionus*, Przewalski horse), it has been reported that keeping males housed singly or only with one female may result in severe aggression and even death upon reintroduction to a herd.

Understanding mechanisms of mate choice and patterns of individual variation is also crucial for successful breeding of many species. For example, a survey on random pairings of tree shrews revealed that only 20 per cent resulted in successful pair formation. Mate incompatibility and rejection of apparently reproductively healthy individuals presents a major problem for breeding facilities housing small numbers of individuals of a particular species. Mating recommendations in captive-breeding Progra-mmes based solely on genetic and demographic considerations may therefore conflict with natural mating strategies and may lead to unproductive management decisions.

PARENTAL CARE

Management practices that provide the appropriate environmental and social requirements for natural parental behaviour are critical for successful propagation of captive mammals. In particular, the survivorship of young may be affected when parental care strategies are not carefully considered for each species. Although maternal care is universal among mammals, the occurrence and extent of paternal and alloparental care varies substantially among species. Many mammalian females require complete isolation from conspecifics and seclusion from potentially stressful events around the time of birth and early rearing.

This appears to he particularly important for species in which only maternal care is provided. Maternal neglect, abandonment, or cannibalism of young has historically been one of the most common causes of death among captive-born mammals. Indeed, it seems that a large portion of unusually high infant mortality rates reported for several captive-bred mammal species may be attributed to parental care problems. In tree shrews, for example, it has been documented that about 50 per cent of young are lost due to maternal neglect or cannibalism. Changes in environmental factors such as the provision for more than one den site can provide solutions to maternal care problems in some species. For example, in the wild, cheetahs and several other carnivore species move their young when disturbed. Disturbance in captivity, such as a thunderstorm or other unpredictable events, may cause a similar reaction in the female. If only one den site is provided, cubs may be carried around excessively and eventually die of hypothermia or bite wounds. Behavioural studies on tree shrews showed that females will not share a nest with their young, but rather visit them and then return to their own den. Provision of additional nest-boxes in captivity therefore led to an increase in breeding success. In several ungulates where mothers spend a large amount of time away from their young in the wild appropriate hiding places are necessary to facilitate this separation of mother and infant.

Reportedly, injuries and maternal care problems may result if constant visual contact of mother and infant is enforced due to inappropriate enclosure design. Maternal incompetence has been reported for a number of species and may result from lack of experience. Recommendations for captive management have been provided based on these findings. For example, in bottlenose dolphins *Tursiops truncatus,* it has been suggested that first-time mothers should be kept with experienced females for enhanced rearing success and in gorillas and chimpanzees considerable effort has been undertaken to teach mothering skills to females.

Overall, parental care behaviour appears to be a particularly labile function and can be disrupted easily even at low levels of stress when other aspects of reproductive function are normal. Hand rearing of captive-born offspring has often been employed to bypass some parental care problems encountered in many species. For example, Nadler reported that more than 80 per cent of captive-born gorillas were hand reared. Similarly, in maned wolves *Chrysocyon brachyurus,* the majority of offspring have been hand reared because captive females rarely care for their young. However, this form of intervention may further complicate the longer-term prospects for obtaining a self-sustaining captive population. In some species, where paternal care or alloparental care are part of the natural care pattern, the presence of conspecifics may be necessary for successful rearing. For example, the presence of the male seems to be essential for successful captive rearing in bush dogs. Survival of young was also highest for green acouchi *Myoprocta pratti* when male and female were present. Dudley showed that pups of mice *Peromyscus califomicus parasiticus* were heavier during early development when the male was present. In a number of mammal species females rear their young cooperatively or communally. Colobine monkeys, for example, often transfer their young soon after birth to other conspecific females for prolonged periods of time. Alloparental care is also common in canids, rodents, bats, dolphins, and many other species. Such helping experiences may be important for the adequate development of mothering skills and may be crucial for reproductive success.

SOCIOSEXUAL COMPETENCE

Transmission of intraspecific, learned behaviours from one individual to another is an essential part of early development in mammals. Experience with captive breeding of many mammals suggests that behavioural anomalies can result from rearing individuals in socially deficient environments. Social deprivation during early development is an underlying cause for breeding and parental care problems in captive-bred species. For example, social deprivation of infant gorillas has led to deficits in social, sexual, and parental behaviour, and these effects appear to be difficult, if not impossible, to reverse. Social access during the first year of life is associated with higher reproductive success, and mother-reared females may be reproductively more successful than handreared ones. Hand rearing has been documented to result in behavioural deficiencies in adult individuals of several mammals. This practice may lead to a vicious cycle by producing more behaviourally deficient adults, which may either not breed or not care for young and therefore necessitate further hand rearing of offspring.

Furthermore, even in cases where maternal care is provided, premature removal of young from their mother, peers, or family group may have a negative impact on sexual and social competence because development may not be completed at the time of removal. Deficiencies in motor patterns may also result from inadequate social exposure during developmental stages. Improper orientation and mounting by males and failure to assume the lordosis posture in females are some of the problems observed in a variety of mammals. In addition to the social environment, the complexity of the physical environment can have far-reaching effects on the developing animal.

Both neural and muscle tissue have certain "windows of vulnerability" during which they are particularly sensitive to environmental disruption. Appropriate stimuli are necessary to facilitate optimal development. Laboratory research, mainly on rodents, found a number of morphological, physiological, and behavioural consequences as a result of rearing in more complex environments. Some of these consequences include changes in brain morphology and differences in size, number, and complexity of neural synapses.

Important behavioural changes include greater behavioural diversity and complexity when exploring objects. Such differences in exploratory behaviour appear to reflect functional differences in learning and may be important for an animal's ability to behave adaptively when faced with a novel situation. Lack of environmental complexity may lead to severe behavioural deficiencies and development of rigid and abnormal behaviour patterns (stereotypic behaviours). For example, when reared in enriched environments, black-footed ferrets were able to kill hamsters more efficiently than individuals reared in less complex environments. As a consequence, it has been suggested that animals reared in deficient environments may also be less able to cope with novel and stressful situations such as mate introduction and caring for young.

STRENGTHS AND WEAKNESSES OF CAPTIVE BREEDING

STRENTHS

Behavioural research can be noninvasive, of relatively low cost, and can be carried out by trained zoo personnel, making it an ideal first approach to management problems. The provision of an environment in which natural (or close to natural) behaviours can be exhibited enhances the educational and research potential of a captive population. In addition, this practice is likely to decelerate adaptation to the captive environment and to increase breeding success, which in turn will permit the establishment of self-sustaining captive populations. Insufficient knowledge about husbandry and behavioural requirements are frequently the main impediments to successful captive propagation for many mammals.

Systematic behavioural studies on wild and captive specimens can therefore allow us to provide the appropriate environment for a species to breed naturally (unassisted) in captivity. This will also facilitate the development of behavioural competence and preservation of the behavioural repertoire associated with breeding and courtship. Learned and culturally transmitted behaviours are frequently an important part of early development in most mammal species. Although some behaviours may require little experience to be exhibited, others need to be acquired during various stages of early development in more or less complex social groupings in interactions with parents, siblings, and other conspecifics. Such behaviours may be lost permanently if no provisions are made for their preservation.

Behavioural research can help identify problems and allow us to provide the appropriate social and physical environment for successful development. Finally, behavioural studies are necessary for reintroduction efforts. Appropriate social structure for release, foraging skills, predator avoidance, orientation, and dispersal are some of the important behavioural strategies that need to be examined carefully before any reintroduction attempts to enable the provision of adequate prerelease training.

WEAKNESSES

Behaviours may vary widely over time and between individuals. A trade-off between time spent investigating individuals in one area and overall number of individuals is necessary. Therefore, results are frequently based on relatively small sample sizes and on data obtained from one facility. However, innovative research approaches employing a variety of techniques, such as keeper questionnaires, remote sensing, and video recording, in combination with direct behavioural observations, can provide data from multiple institutions over the same time period. Nevertheless, great care is required to standardize behavioural studies across individual researchers and facilities. The high variability in behaviour makes it difficult for behavioural studies to generate simple guidelines comparable to demographic and genetic guidelines for captive breeding. Instead, a species-specific approach will have to be applied, and even within a species, individual variation and individuals' needs may have to be considered to achieve reliable reproduction in captivity.

For example, individual cheetahs in captivity showed quantifiable behavioural variation with regard to fearfulness and other characteristics. A correlation was found between the degree of fearfulness and breeding status. Nonbreeders scored significantly higher than breeders on the fear component. Studies on domestic dogs and laboratory animals have shown a high heritability of fearfulness and selection may therefore occur for such traits in captivity. Temperamental differences and patterns in individual behavioural variation may therefore require consideration to counteract the adaptation to a specific captive environment. The captive environment may need to be adjusted accordingly to allow different types of individuals to reproduce successfully in captivity. Behavioural studies are time intensive and often do not produce conclusive short-term results.

Alone they may not provide all the information necessary to understand behaviourally based captive management problems such as mate choice, reproductive suppression, stressrelated problems; an interdisciplinary approach may be advisable in most cases.

- *First,* Appropriate enclosure design and furnishing are often established based on anecdotal and descriptive information and frequently on scant field data. More systematic investigation of the effects of various enclosure features on reproductive behaviour and parental care are urgently needed for numerous captive-held species.

- *Second*: The dynamics of interspecific competition and the trade-off of close coexistence among different species in the wild and their possible ramifications for captive management have rarely been systematically examined and are in need of further investigation. In addition, the effect of housing prey and predators in close proximity needs further attention.

  Considering the preservation of naturally occurring behaviour patterns, it may be beneficial for prey species to have some degree of exposure to predators as long as ample hiding spaces and room for retreat are provided.
- *Third*: Studies on wild and captive populations of many species are necessary to identify species-specific foraging patterns and to investigate their relationship to other important behavioural variables, such as reproductive behaviour and parental care.
- *Fourth*: There are many captive species in which problems associated with mating, incompatibility, and unknown reproductive behaviour are regarded as a major impediment to captive propagation. Although behavioural studies have provided solutions for some of these breeding problems, an understanding of underlying mechanisms and causes has not been achieved for many captive species, and behavioural-physiological research is urgently needed in this area.
- *Fifth*: Establishing time budgets and behavioural repertoires for wild populations is needed to provide a baseline for comparison to behaviours and time budgets found in captivity and to allow identification of abnormal behaviour patterns.
- *Sixth*: Combined genetic and behavioural studies

identifying parentage in wild populations can help to establish the true mating systems of a species, which may in some cases differ from previous findings based on behavioural observations alone.The above recommendations are based on the literature and are intended to provide a few salient examples of necessary future research. Many more could certainly be added to this list. In addition, four general recommendations can be made.

- *First*: information from descriptive and anecdotal accounts on "do's and don't's" in captive breeding needs to be synthesized for each species and used to generate testable hypotheses.
- *Second*: the development of standardized methods and training procedures for behavioural research in captive studies is necessary to allow for broader application of results. These methods should then be applied across facilities and countries. In particular, multi-institutional studies and innovative methodological approaches to the study of behaviour should be encouraged.
- *Third*: integration of behavioural research with other research disciplines is becoming increasingly necessary. Interdisciplinary studies, for example, combining behaviour and physiology (increasingly feasible through the recently developed noninvasive techniques of fecal, urinary, and saliva steroid monitoring), will allow us to test hypotheses about complex aspects of behaviour and the interplay of behaviour, physiology, and various environmental factors.

Finally, guidelines and provisions concerning behavioural management of endangered species need to be an official part of captive-breeding management plans. Consultation with behavioural ecologists should occur at the beginning of any planned breeding Programme.

7

Physiology of Insects

All organisms need to communicate with the biotic and abiotic components of their environments. Organisms also communicate within themselves. We saw during our discussions of homeostatic mechanisms and physiological adaptations that insects need to register conditions and changes within themselves and in their environments, and then respond to the conditions.

The physiological basis of homeostatic mechanisms is that an organism can somehow measure aspects of the parameters that are regulated. For examples, changes in hemolymph salt concentrations and thoracic temperature need to be registered before corrective action can be initiated. Insects also respond to changes in their environments. Mechanisms of entering diapause seem a particularly good example. Here, insects register token stimuli, most often changes in photoperiod. These token stimuli are translated in neuroendocrine events, which in turn are translated into physiological changes.

Nervous systems play crucial roles in perceiving environmental conditions, integrating environmental signals into meaningful information, and then producing responses in the forms of internal, physiological events. In another, speedier set of conditions, try imagining how an insect would fly if it could not see where it was going, or could not determine orientation with respect to its horizon. Within a flying insect, again, try imagining losing control of the small, direct flight muscles that control pronation of wings, or loss of coordination of the major flight muscles. Also consider various sorts of homeostasis during flight: internal temperatures are measured, then hemolymph flows adjusted accordingly; available fuels in hemolymph are oxidized, then replaced from stores quite distant from flight muscles. It would be difficult to regard any area of insect biology that does not depend importantly upon nervous systems. Understanding insect nervous systems also has a certain practical significance because most insecticides work by poisoning one or another area of nerve function. The nervous system is made of neurons. Every tissue is composed of cells. Fat bodies are mostly made of trophocyctes. Muscles are composed of muscle cells. Similarly, nervous systems are composed of nerve cells, which are also called neurons. The general structures of nerve cells. Nerve cells have cell bodies, called perikarya (pleural of parikaron). Most major organelles, including cell nuclei, endoplasmic reticula, Golgi bodies, mitochondria and so forth are found in perikarya.

Long projections known as axons (pleural of axon) extend from cell bodies; axons carry nerve impulses to their ends, often called terminal arborizations. These arborizations are typically juxtaposed very closely to other nerves or to muscles. At these points, called synapses, electrical impulses are transduced into chemical signals, which are received at membranes of other nerves or muscles. Nerve cells receive stimuli at their dendrites (pleural of dendrite). Dendrites can be found directly on perikarya or at ends of axons. We consider three kinds of neurons.

The upper drawing shows a monopolar neuron, which have a single projection from parakarya. This single projection branches so that there is a dendrite at one end and terminal aborization at the other. Most insect neurons are monopolar. The middle sketch shows a bipolar neuron; these neurons are commonly associated with sense cells, and feature a dendrite at a distal end and a terminal aborization at the other. This arrangement is suited for direct transmission of environmental stimuli to the central nervous system. Multipolar neurons have several axons projecting from perikarya; this cell type occurs in some ganglia, and is also connected with stretch receptors. Information from outside of an insect, such as light or mechanical stimulus, is most likely conducted from peripheral sense cells directly to central nervous system ganglia by axons without synaptic interruptions. These axons are called afferent axons. Other axons are involved in conducting motor commands from central nervous system ganglia, wherein lie the cell bodies, to muscles; these axons are called efferent or motor axons. Most afferent neurons synapse with other neurons that in turn synapse with still other neurons or else with efferent neurons. The neurons that form connections between afferent and efferent neurons are called interneurones. Nerve fibres are generally aggregates of several or many neurons.

CENTRAL NERVOUS SYSTEM

Unlike vertebrates, in which a single, relatively large brain is encased in a skull, central nervous systems of insects are composed of variously distributed ganglia. The major association centre is called the brain. The other important ganglia are subesophageal ganglion, frontal ganglion, up to three thoracic ganglia and up to eight abdominal ganglia. Three major regions of brains are recognized: protocerebrum, deutocerebrum and tritocerebrum. The protocerebrum is the most complex part of insect brains; it has two lobes, and is continuous with the optic lobes. The deutocerebrum contains antennal lobes, from which project both afferent and efferent neurons.

The tritocerebrum is a small, bilobed ganglion. Major nerve tracts connect the tritocerebrum to other ganglia. The tritocerebrum is linked to subesophageal ganglion by circumesophageal connectives and to the frontal ganglion by frontal commisure. This part of the brain has efferent and afferent connectives to certain mouthparts. The subesophageal ganglion is the first ganglion in the ventral nerve cord. This ganglion has motor and sensory connections to salivary glands, mouthparts, and neck. Most insects have three thoracic ganglia, each with connectives to muscles and sensilla. Abdominal ganglia have connections to muscles, though usually fewer than thoracic ganglia. Numbers of ganglia vary among groups of insects. The thoracic and abdominal ganglia are fused into a single compound ganglion in houseflies. This is an example of extreme fusion. The least fusion is shown in a thysanuran. The last abdominal ganglion is always a compound structure, derived from the ganglia of the last four segments. The structure of a ganglion. The central nervous system is completely invested in what is called the nerve sheath, the most outer part of which is a non-cellular neural lamella and the inner part of which is called perineurium, composed of a single layer of cells. The perineurium is thought to produce and secrete the mucoprotein and mucopoly-*saccharide neural lamella. The perineurium serves as a blood-brain barrier, and as such it regulates the chemical environment of nerve cells and transports nutrients and other materials between hemolymph and nerve cells.

The centres of ganglia are made up of axons and fibres of afferent, efferent and interneurones. These centres specifically do not contain cell bodies, and they are called neuropile. Within neuropiles, some axons and fibres are orientated such that they form fibre tracts. Cell bodies, or perikarya, are found around the periphery of ganglia. The inter ganglionic connectives are similarly arranged, with neural lamella and perineurium forming a sheath around the tracts and giant fibres running along in the middle. Within neuropiles and connectives, each neuron is surrounded by glial cells which form a protective barrier around nerve cells. Glial cells can have a number of arrangements, and are often folded around axons, in which case they are called mesaxon. Glial cells probably insulate nerve processes from each other to reduce uncontrolled electrical noise or "cross-talk" from one nerve to another. Glial folds do not occur around synapses.

TRANSMISSION

The important features include presynaptic vesicles, many filled with transmitter substance, usually acetylcholine in nerve-nerve synapses, and L-glutamate in nerve-muscle junctions. Gamma-aminobutyric acid (GABA) is the usual transmitter chemical in inhibitory junctions. Other features are presynaptic membrane, and postsynaptic membranes. Finally two proteins are important, acetylcholine esterase and acetylcholine receptor sites. This drawing shows the synaptic cleft, which separates pre- and postsynaptic membranes by 200 to 500 angstrom units.

Contemporary wisdom holds that depolarization of presynaptic membrane leads to fusion of presynaptic vesicles with membrane, and release of acetylcholine by a process of exocytosis. Vesicles contain on the order of a few thousand molecules of transmitter; each depolarization causes about 100 or so vesicles fuse with presynaptic membranes and release their molecules. Acetylcholine molecules move randomly as in free solution within synaptic clefts, and it is about equally possible for any given molecule to meet acetylcholine esterase or acetylcholine receptor sites. The enzyme hydrolyzes acetylcholine to acetate and choline. The acetylcholine molecules that bind to their receptor sites eventually cause a depolarization of the postsynaptic membrane called excitatory postsynaptic potential (EPSP), which is transmitted to the axon. Axon depolarizations and continuation of the nerve impulse follow.

INSECT VISION

Vision is the perception of light. Insects perceive light through three classes of sensory organs. Most adult insects have a single pair of compound eyes. Larvae of hemimetabolous insects and most adults have usually three simple eyes called ocelli. These are typically located on the dorsum of the head capsule, and they are sometimes called dorsal ocelli. Larvae of holometabolous insects do not have compound eyes. These insects perceive forms, to a limited extent through stemmata located on either side of the head.

COMPOUND EYES

The compound eyes are prominent features of adult insects. The eyes occupy a fairly large portion of the surface of insect heads, and they facilitate a rather wide field of vision. Compounds eyes are not present in all insects. They are reduced or absent in parasitic forms, many soil insects, and in some species that live in very dark places, such as caves. The basic unit of compound eyes is the ommatidium (singular of ommatidia). Ommatidia vary in size and number among insect groups. At one extreme, dragonflies have thousands of ommatidia in each compound eye; many insects have far fewer, and at the other extreme, the workers of the ant species *Pomera punctatissima* have only one ommatidium in each eye. The sizes of ommatidia vary from about 5 to 40 microns in diameter. Sizes vary among species and even within a single compound eye. In some dragonflies, for example, the dorsal units are considerably larger than the ventral ones.

The structure of a single ommatidium. The ommatidium rests on a basement membrane. The height of the eye comes from the long retinula cells and secondary pigment cells. The corneagen cells (sometimes called primary pigment cells) rests atop the long retinula and secondary pigment cells. The crystalline cone lies within the corneagen cells. The surface of the ommatidium is covered with the corneal lens. The corneal lens is a specialized part of insect cuticle. It is secreted by the corneagen cells, which are specialized epidermal cells. The crystalline cone is a clear, intracellular structure synthesized within Sempter cells. The primary pigment cells surround the Sempter cells. Not all compounds eyes feature a crystalline cone; those that do are called eucone eyes. The corneal lens and crystalline cone make up the optical part of ommatidia. The sensory elements are just under the optical apparatus. The retinula cells are long neurons. Part of each retinula cell is a specialized area known as a rhabdomere. A nerve axon from each retinula cell projects through the basement membrane into the optic nerve. Ommatidia are functionally isolated because the retinula cells are surrounded by the secondary pigment cells.

Compound eyes are commonly categorized as either apposition eyes or superposition eyes. There is no separation between the corneal layer and the photoreceptors in apposition eyes. There is a clear space between the two units in superpositions eyes. Some authors use the expression "clear zone eyes" for the superposition eyes. Superposition eyes are generally found in crepuscular and nocturnal insects, while apposition eyes occur in diurnal insects. The crystalline cone couples the lens and photoreceptors in apposition eyes. Some higher Diptera have pseudocone eyes.

Crystalline cones do not occur in pseudocone eyes, and optical coupling is by means of a gelatinous substance that is contained in a two-celled structure. Still another arrangement is found in some apposition eyes that lack solid cones or gelatinous pseudocones. In these acone eyes four flat transparent cells are found in place of the cones. In superposition eyes the space between the corneal lines and the photoreceptors is traversed by crystalline tracts. These are thought to act as wave-guides that direct light to the photoreceptors. Tracheal tapeta are known from several Lepidoptera. Small tracheae run parallel with the longer cells of the ommatidia. It has been suggested that these structures function as interference reflector filters.

Light sensitivity of insect eyes varies according to the state of light- adaptation or dark-adaptation of the eyes. In bright light, the eye is less sensitive to light, and the eye is regarded as light-adapted. Maximum sensitivity occurs in darker conditions, when the eye is fully dark-adapted. Some insect eyes can change light sensitivity by about three orders of magnitude within a few minutes of changing light conditions. There are a number of mechanisms of adaptation to light and dark conditions. One mechanism relates to the biochemistry of vision. The visual pigment is broken down by interaction with light. In daylight the breakdown rate can equal or exceed the replacement rate. This leads to decreased light sensitivity, and partly explains light adaptation. In the dark, the visual pigment accumulates, and the insect becomes dark adapted.

A second mechanism of adaptation is the movement of screening pigments. Screening pigments are located in the retinula cells. In light-adapted eyes granules of screening pigments surround the rhabdom. These pigments absorb light, and they have the effect of optically isolating individual ommatidia. The endoplasmic reticulum of dark-adapted eyes forms large, clear vesicles. This forms a clear space around the rhabdom, so that a greater amount of light falls upon the rhabdom. Vision involves the transduction of light energy into a bioelectric signal within the nervous system. The first events in this process take place in the retinula cells. The fine structure of rhabdomeres consists of thousands of closely packed tubules. These tubules are about 500 A in diameter and one micron long. They are aligned at right angles to the long axis of the rhabdom. The visual pigments occur mainly in these rhabdomeric microvilli. It has been suggested that the small diameter of each microvillus inhibits free rotation of visual pigments. This specific orientation may be the molecular basis of insects' sensitivity to polarized light.

Photobiological processes occur in a narrow band of the electromagnetic spectrum between 300 and 700 nm. Photons in this region of the spectrum have enough energy for photochemical interactions, but not enough energy to disrupt macromolecules. Visual pigments initiate vision by absorbing light in this region. These pigments are a class of membrane bound proteins known as opsins that are conjugated with a chromophore. Visual pigments whose chromophore is retinal are called rhodopsins. The visual pigments of all invertebrates, including insects, crustaceans and squids, are all rhodopsins. There are a number of geometric isomers of retinal. The 11-cis isomer is the chromophore of native rhodopsin. Free retinal absorbs light strongly at about 380 nm. When 11-cis-retinal is linked to opsin to form rhodopsin, the main absorption peak is shifted. The maximum absorption values of different rhodopsins range from about 345 nm (ultraviolet) to as high as 610 nm (red).

The absorbing properties of any given rhodopsin is related to the disposition of charged opsin groups around the chromophore. These interactions between the side-groups of the amino acids within the opsins and the chromophores are thought to modify the p electron orbital of the chromophores, and thereby modify the absorbance spectra of rhodopsins. When the chromophore of rhodopsin absorbs a photon, the chromophore undergoes an isomerization from the 11-cis to the all-trans configuration. The rhodopsins of insects and other invertebrates differ from vertebrate rhodopsins on the point of the chemical events that follow the isomerization step. Vertebrate rhodopsins go through a series of spectrally distinct intermediates. The process is called bleaching because the absorption shifts from visible to ultraviolet wavelengths, which we can not see. The vertebrate process can lead to hydrolysis of the opsin and chromophore. Invertebrate rhodopsins generally do not bleach. Light transforms the rhodopsins to a stable intermediate. The isomerized 11-cis chromophore stays in place in the microvilli. The stable intermediates have spectral properties similar to metarhodopsin I of vertebrates, which has earned them the term metarhodopsins.

The mechanism of transducing light-induced chemical changes in rhodopsins into electrical signals remains the major unresolved issue in understanding the physiology of light receptors. Several milliseconds are required before the absorption of light can be recorded as electrical activity in the receptor cell membrane. Certainly, the phototransduction event occurs during the transition from rhodopsin to metarhodopsin. We can not say much more about the detailed mechanisms of transducing light into bioelectrical signals. A key requirement in the chemistry of vision is that metarhodopsin, or the bleached pigment of vertebrates, must be reconverted to rhodopsins. In the eyes of vertebrates, this regeneration is the product of a complex biochemical pathway. The pathway involves isomerization back to the 11-cis configuration and assembly with the opsins. Again, invertebrates differ from vertebrates on this aspect of photochemistry. Light is absorbed by rhodopsin and metarhodopsin in invertebrates.

Absorption of light transforms metarhodopsin back to rhodopsin. The pigments exist in two stable conformations, and light causes them to switch back and fro from one state to the other. This regeneration is essential because conformational changes in rhodopsin, but not in metarhopsin, leads to phototransduction. Experiments on insect behaviour indicate that many insect species can distinguish colours. Colour vision depends on the ability to discriminate between light of different wavelengths. Such discrimination is possible because compound eyes have photopigments with differing spectral properties. Many insects have two rhodopsins, one with maximum absorption in ultraviolet wavelengths and one with maximum absorption in green wavelengths. Some species have a third pigment that absorbs maximally in the blue region. A few Lepidoptera have four visual pigments.

DORSAL OCELLI

Dorsal ocelli occur in larvae of hemimetabolous insects and in nearly all adults. Here is the structure of a dorsal ocellus. Although there are variations in structure, a typical ocellus has a single lens that is usually rather thickened. Most ocelli feature a large number, often hundreds, of retinula cells. Rhabdomeres of several cells combine to form a number of rhabdoms. Axons project from the retinula cells through the basement membrane and they end in a synaptic plexus behind the eye. Light causes a sustained depolarization of the retinula cells, however, the biological roles of ocelli remain unknown. It is generally agreed that ocelli allow only poor, if any, perception of form. In some orthopterans the ocelli are active in orientation to a light source.

STEMMATA

The larvae of holometabolous insects have stemmata on the sides of the head capsule. Many larvae have a single stemma (singular of stemmata) on either side, but the number can be as many as six on a side. Here is a sketch of a section through a stemma. Each stemma has a cuticular lens and a crystalline cone distal to the sense cells. Stemmata generally do not produce clear images, but most caterpillars can discriminate some shapes and they can orientate themselves with respect to boundaries.

BIOCHEMISTRY OF INSECT BODY

HOMEOSTATIC MECHANISM

Homeostatic mechanisms maintains stable conditions

with respect to any given physiological parameter. Consider homeostasis of thoracic temperature during flight; of hemolymph energy metabolite concentrations; of water balance in terrestrial insects; of osmotic balance; of symbiotic organisms. Two aspects of physiology are implied in discussions of homeostatic mechanisms: first, organisms can register changes in an homeostatic parameter, and second, features of the parameter can be adjusted. In homeostasis of thoracic temperature, it is clear that thoracic temperature is monitored during flight (even if we do not understand the measuring system), and it is clear hemolymph circulation patterns are adjusted in response to changes in thoracic temperature.We introduce a discussion of the insect fat body in terms of homeostasis for a couple of reasons.

In most insects the fat body serves as a storage depot for food reserves, and sometimes for storage excretion. Lipid reserves are often accumulated in massive quantities in this organ, so that it looks like a body, or organ, of fat. The term fat body was coined during early days of studying insect morphology, and it derives from the fatty appearance of the tissue. This term is deeply entrenched in the literature of our field, so much so that it would be a futile exercise to try working towards a more descriptive expression for this organ. The alternative is to expand our conception of what the term fat body represents. In addition to its important roles as a storage depot, the fat body of insects functions as a key centre of metabolism and biochemistry. As a metabolic and biochemical centre, the biological significance of fat body is its ability to maintain a balance between resources and requirements during the many phases of an insect's life. On one side of the balance, during times of feeding fat bodies biosynthesize and accumulate not only lipid reserves, but also carbohydrates, amino acids, proteins and other metabolites.

On the other side, fat bodies respond to physiological and biochemical needs in a number of ways, including very high rates of protein biosynthesis, formation and release of trehalose, release of lipids, detoxification of nitrogenous waste products, and biosynthesis of hormones. Many of the responses to physiological needs occur on a relatively large scale, and they can have substantial impact on insect biology. Very often fat body metabolic functions are regulated by various endocrinological mechanisms.

STRUCTURES OF FAT BODIES

The morphology of fat bodies varies considerably in various insect groups. In many species it is a loose aggregation of cells that may be unevenly distributed among all body segments, including the brain case. Fat body cells often surround the major organ systems in insects, including reproductive tissues, alimentary canals, thoracic muscles, and elements of central nervous systems. This close association of fat body cells and other tissues may facilitate exchanges of macromolecules. Fat bodies may also protect organs from physical damage by providing packing. In other species fat bodies are compact cell masses invested in a membranous covering such that they easily can be removed from the insect intact.

Structures of fat bodies may vary according to life stage, sex and reproductive status. Fat bodies can have different functions between sexes, and functions also can differ among life stages. In some species they undergo histological breakdown and reformation during papal stages, though in most species they do not appear to do so. Fat bodies of all insects are in intimate contact with circulating hemolymph, which is consistent with movements of molecules between the two compartments. The vast majority of the cells in fat bodies are called trophocytes, and it is in these cells that most fat body functions are carried out. Trophocytes are very similar to some hemocytes; the two cell types may be quite closely related. In the homopteran Aleyrodes, for example, fat body cells float individually in hemolymph and can not be clearly distinguished from hemocytes.

Other fat body cell types include urate cells. These cells are associated with storage excretion, in which uric acid crystals are accumulated rather than excreted. Fat bodies of most insect species do not include urate cells. For the sake of complexity, uric acid is stored in trophocytes of some species. Fat bodies of many species also contain mycetocytes, which are involved in housing symbiotic micro-organisms. Many of these symbiants play crucial roles in insect nutrition by providing certain vitamins and other essential nutrients to insects that are not able to acquire these nutrients in the diet.

PROTEIN BIOSYNTHESIS IN FAT BODY

Many of the proteins that are crucial in the lives of insects are biosynthesized in fat bodies. Protein synthesis is often under control of juvenile hormone (JH). One of the most well-studied examples of endocrine regulated protein synthesis in fat bodies is the formation of vitellogenins by adult females. Vitellogenins are the yolk proteins that are accumulated during egg maturation. Fat bodies biosynthesize many other proteins as well, including virtually all of the major proteins found in the hemolymph of larval insects. There are also many examples of specialized proteins in insects. Larvae of the midge *Chironomus* biosynthesize hemoglobins, heme-containing proteins important in storage of oxygen. Hemoglobins can make up more than 80 per cent of the hemolymph proteins in these larvae. The lipoproteins that function in hemolymph transport of diacylglycerol are synthesized in fat bodies. There are many other examples, all bearing on the point that protein synthesis in fat bodies is a central issue in insect biology.

We review the biochemistry of protein synthesis by way of appreciating the importance of this topic. First we recall the expression originally put forth by Francis Crick, and later articulated by Charlie Brown as the central dogma of molecular biology: DNA Makes RNA Makes Protein. As an aside, a group of my biochemical colleagues at the University of Nevada, Reno and myself once entered a biochemical T-shirt logo contest. We added the expression makes lipids to the expression of the central dogma of molecular biology. We did not win. The transcription process produces a special form of RNA called messenger RNA, or mRNA. Messengers go places, and in this case mRNA makes its way from the nucleus to the cytosol where it joins with ribosomes. We want to see that the mRNA is joined with two ribosomal subunits, a small and a large subunit. The flow of information is from the nucleus to the cytosol. This is followed by a series of steps that include translation of the genetic information borne on the mRNA strand into a polypeptide chain, or string of amino acids.

The key here is that each of the amino acids can be attached to a specific RNA molecule known as transfer RNA, or tRNA. One portion of tRNA molecules feature an anticodon that matches up to codons on the mRNA strand. Ribosomes have sites for two tRNA molecules, one called a peptidyl site and the other called an aminoacyl site. While two tRNA molecules are juxtaposed within the ribosome a peptide bond is formed between the amino acids attached to each tRNA. At this point the tRNA molecule in the peptidyl site is released and the ribosome scoots one step along the mRNA strand, thereby shifting the tRNA molecule in the aminoacyl site to the peptidyl site and opening a space for a new tRNA molecule. The cycle goes on with addition of the new tRNA molecule, formation of another peptide bond, release of a free tRNA molecule and another ribosomal shift. A copy of the universal genetic code is appended to the end of this column. Please note, for example, that the AGG in the mRNA strand shown above specifies the amino acid arginine.

Protein synthesis begins with transcription of a DNA strand within a cell nucleus. JH expresses its effects on protein synthesis within fat body cells by interacting with DNA. JH receptor proteins are found within the cytosol of cells. JH forms a complex with its specific receptor, then the hormone-receptor complex moves into the nucleus. It is the JH-receptor complex that interacts with DNA.

CARBOHYDRATE METABOLISM IN FAT BODY

The natural foods of many insects include substantial amounts of carbohydrates, or sugars, either in the form of simple sugars or as polymers. Following absorption across midgut epithelium, sugars are removed from circulating hemolymph into fat bodies. Sugars are assembled into large macromolecules of glycogen and stored within fat bodies (try to imagine a discussion on the physiology of the sugar body). The polymerization of simple sugars into glycogen is called glycogen synthesis, and is carried out enzymatically by glycogen synthetase. It has been suggested that glycogen synthetase may be under control of a sort of hypoglycemic hormone. There are a number of lines of evidence for factors that reduce hemolymph sugar levels, but the specific points of control remain unknown.

But there is also evidence for endocrine regulation of carbohydrate mobilization. Various names have been applied to hormones that increase hemolymph sugar levels, including hypertrehalosemic factor and glucogon-like hormone. Most recently people have been using the term trehalogon by analogy to the mammalian hormone Glucogon. Insect endocrinologists a bad name, but there we have it. Trehalogon is thought to stimulate fat bodies to release trehalose into circulating hemolymph. Releasing trehalose into hemolymph is a multi-step process that begins with hydrolysis of individual glucose residues from glycogen.

This enzyme yields molecules of glucose-1-phosphate, which are taken into two pathways, one is a mutase step to glucose-6-phosphate and the other is a phosphorylase step to uridine diphosphoglucose. These two glucose derivatives are combined into trehalose-6-phosphate by trehalose-6-phosphate synthetase. A phosphatase step yields the free trehalose molecule. Hence, trehalose synthesis requires hydrolysis, derivatization and assembly of two glucose residues.

Trehalose synthesis is regulated by controlling the activity of phosphorylase, the enzyme at the beginning of the pathway. There appears to be two forms of phosphorylase, phosphorylase a, the active form and phosphorylase b, the inactive form. Both of these forms exist in fat bodies, and the specific ratio of one form to the other depends upon the physiological requirements of the whole organism. Physiological requirements are expressed by circulating titres of trehalogon, the hormone active in mobilizing fat body trehalose. It contains a slight error, shown in the phosphorylase step.

METABOLISM OF LIPIDS IN FAT BODY

We have seen that lipid reserves are stored in fat bodies as triacylglycerol. Triacylglycerol is synthesized in fat body from excess sugar and from the carbon skeletons of certain amino acids by way of the ubiquitous enzyme system fatty acid synthetase.

Newly formed fatty acids are linked to glycerol backbones by acyltransferases. Lipid is released from fat body in the form of 1,2-diacylglycerol (DAG), rather than as 1,3-diacylglycerol or a mixture of 1,2-DAG and 1,3-DAG. 1,2-DAG may be formed directly by a lipase specific to the 3-position fatty acid, or it may be formed by breakdown to 2-monoacylglycerol followed by stereospecific reacylation to 1,2-DAG. Release of 1,2-DAG is stimulated by adipokinetic hormone, but the particular enzyme target is not yet known.

DETOXIFICATION REACTIONS

The details of uric acid synthesis will be covered in the section on nitrogen metabolism and excretion. The point to emphasize here is that the fat body is the central organ of nitrogen metabolism. The storage of uric acid in urate cells and in trophocytes by some insects has been covered. One of the great problems in crop protection by application of chemical pesticides is the natural resistance to these poisons that insects rapidly develop.

Two broad mechanisms of resistance are enzymic breakdown steps, one by esterase and another by mixed-function oxidases. Under these systems, the pesticide molecules are simply metabolized into harmless forms before they can carry out their damaging effects. Most of these detoxification enzymes are formed in fat bodies and function in circulating hemolymph as well as in fat body cells.

BIOSYNTHESIS OF AMINO ACIDS IN FAT BODY

The non-essential amino acids can be formed using carbons from various intermediates in carbohydrate metabolism coupled with transaminations from preexisting amino acids. The fat body is the major site of amino acid synthesis.

HORMONE IS SYNTHESIZED IN FAT BODY

Most homeostatic hormones, such as diuretic hormone, adipokinetic hormone and trehalogon are released from elements of corpus cardiacum or from elements of neurosecretory cells. There is an example of a hormone that is produced and released from the fat body. About 40 hours after taking a blood meal, adult female mosquitoes produce what is tentatively called host-seeking behaviour inhibitory factor. This factor is a small peptide, and is produced in fat body and released into circulating hemolymph. The exact site of action remains to be found, but one physiological effect is to switch off specific lactic acid receptors located in antennae. Fat bodies need to be physiologically prepared to synthesize the factor. Another hormone appears to be required because females that have been ovariectomized within three hours after a blood meal do not elaborate the factor.

BIOCHEMICAL ROLES OF VITAMINS

We will use this period to extend our treatment of nutrition by adding details on the biochemical roles of some vitamins. This will also serve as a review of energy metabolism, meant to help prepare for the final examination. We recall that muscle contraction involves interactions between thick myosin filaments and thin actin filaments. These interactions occur by way of cross bridges formed by heads of the myosin molecules. Mg^{++} - dependent Myosin ATPase is responsible for the conformational changes in the orientation of the myosin cross bridges that result in the filaments sliding past on another. The sliding filaments result in a shortening of the Z-band to Z-band distance.

Tracheoles functionally, but not topologically, penetrate deep into rapidly metabolizing muscle fibres such that oxygen can be delivered nearly directly to muscle mitochondria, sometimes called sarcosomes. The key function of the gas exchange is to support energy metabolism during flight, or any other energy-intensive activity. By now we should have an idea of where, morphologically and biochemically, oxygen is consumed and where carbon dioxide is produced. We can revisit energy metabolism by looking back at the chart that shows types and forms of fuels. We can also recall the metabolic pathways that yield energy to support cell work. These include glycolysis, Kreb's Cycle, the Electron Transport Chain and ß-oxidation. A physiologist might say, barring some needling details, our grasp on energy metabolism is virtually unrivalled in the western hemisphere.

The sugars, lipids and protein components that insects and other organisms consume in the course of their feeding activities are called macronutrients, not because they are particularly large or complex molecules, but because they are required in relatively large quantities. In addition to these macronutrients, insects require a host of micronutrients, so called because they are needed in more or less trace levels in the body. Some of these micronutrients are Vitamins, which can be defined by three points. First, they are found in trace to low quantities in the body. Second, they are involved in specific metabolic functions. Third, they are not synthesized in the body. If they are synthesized, the synthetic rates do not support the organismal needs.

Some of the vitamins are directly involved in energy metabolism, acting within the pathways within our line of familiarity. We are going to look at these vitamins, and at the specific metabolic functions they perform. This is one of the areas of biochemistry that integrates common features of many different organisms. These metabolic functions are just as vital to mammals, including human beings. Generally speaking, we recognize two major groups of vitamins - the fat soluble and the water soluble vitamins. The fat soluble group includes vitamins A, D, E, and K. The water soluble group includes the B-complex and vitamin C. We proceed with vitamins we have already seen in action. The structure of riboflavin is shown first. Riboflavin in its vitamin form, as riboflavin, and also in its physiologically functional form in flavin adenine dinucleotide, often abbreviated as FAD. The key feature here is that the vitamin - that is, the trace substance that can not be manufactured within the body - is merely a component of a larger molecule that plays a specific role as a coenzyme. Before we discuss the role of FAD, let us introduce nicotinamide, also shown in its vitamin form and in its physiological form of nicotinamide adenine dinucleotide, similarly abbreviated as NAD.

Both FAD and NAD function as electron carriers. When an electron pair is transferred to these molecules, either one or two hydrogens are also transferred. FAD has accommodation for two hydrogens while NAD can accept one hydrogen. In NAD, then, an electron pair and one hydrogen are transferred, with a second hydrogen released into the medium. This subtle chemical difference is why the reduced form of FAD is written FADH2 while the reduced form of NAD is written NADH + H^+.

These compounds are particularly appropriate electron carriers because they readily accept an electron pair and they also readily donate an electron pair. We saw FAD act in the Kreb's cycle. FAD accepts an electron pair from succinate, becoming FADH2. FADH2 carries the electron pair to the electron transport chain, where it releases its electrons at such position in the chain that two ATPs are subsequently formed. The figure also shows that NAD operates as an electron carrier at three places in the Kreb's cycle, carrying electrons from isocitrate, alpha-ketogluterate and from malate to the electron transport chain. Here the electrons are loaded onto the chain such that three ATPs are formed. These examples illustrate a general feature of co-enzymes: they all provide some chemistry to enzymes that the proteins can not perform on their own.

As coenzymes, FAD and NAD provide the ability to accept and donate electrons. Riboflavin and nicotinamide are necessary components of these co-enzymes, and these components can not be made in the body of insects. For perspective, recall from our biochemistry that NAD acts as a co-enzyme in more than 250 enzyme functions of this sort: carrying electrons from one place to another. Our next water soluble vitamin is B-1 or thiamin, shown in its vitamin form and in its physiologically active form as thiamin pyrophosphate, or what is sometimes called cocarboxylase. This vitamin is quite often referred to as TPP. As a point of history, it was because of the amine group in this compound that accessory food factors were first called vitamines, an abbreviation of vital amines. Specifically, TPP is a coenzyme in the removal of carbon dioxide from various compounds. An important reaction is this one:

$$\text{Pyruvate} \rightarrow \text{Acetyl CoA} + CO_2$$

This reaction is carried out by pyruvate decarboxylase. This reaction removes the carbon dioxide moiety from pyruvate, and it connects the glycolytic pathway to the Kreb's cycle. Returning to the Kreb's cycle, note that carbon dioxide is released at two points in the cycle. The active form of thiamin, TPP, also functions in the release of this second carbon dioxide. We collect the point that some decarboxylations, or carbon dioxide removal reactions, do not depend the activity of this co-enzyme.

We have now seen three vitamins - two function as electron carriers, essentially connecting the Kreb's cycle to the electron transport chain and the third functions in decarboxylation reactions. In one decarboxylation reaction, specifically pyruvate decarboxylase, TPP also helps connect the glycolytic pathway to the Kreb's cycle. The transport of 2-carbon acetyl units, done by Co-enzyme A or CoA, is still another key link between glycolysis and the Kreb's cycle. Recall that oxidation of fatty acids in the ß-oxidation pathway is also connected to the Kreb's cycle by transfer of acetyl CoA. This serves to underscore the importance of vitamin B-3, or pantothenic acid. This is shown in its vitamin form and as a component of Co-enzyme A. These four vitamins - nicotinamide, riboflavin, thiamin and pantothenic acid - are directly involved in energy metabolism; they act as electron carriers, in decarboxylations and as an essential component of CoA. Insects require other water soluble vitamins in their diets, including biotin and folic acid. These do not function specifically in energy metabolism. Nonetheless, we want to briefly consider their roles in cellular metabolism. The chemical structure of biotin was worded out in 1942, and first synthesized in 1943. Biotin had been independently discovered by at least two groups, and it went under a variety of names, including vitamin H and protective factor X. Biotin functions as a carbon dioxide carrier in carbon dioxide fixation reactions. Here is an example:

$$\text{Acetyl CoA} + CO_2 \rightarrow \text{Malonyl CoA} \rightarrow \text{Fatty Acids}$$

Folic acid is actually a complex of several biologically active compounds. The formal term is pteroylglutamic acid, and it is composed of pteroic acid plus one to seven gluatmate residues. Folic acid is required for transferring single carbon units. It is essential in metabolism of some amino acids, and is probably most well known for its critical role in synthesis of thymine. Without folic acid, DNA synthesis is slowed or stopped. In the absence of DNA synthesis, production of some cells, such as red blood cells, is also slowed or stopped. This is particularly important for cells that undergo rapid turnover, such as red blood cells or insect hemocytes. We should be able to describe the general sort of chemistry they provide and be able to give a specific example. We would not want to give much thought to the chemical structures of things or try to memorize many more than about half of the 250 know NAD reactions.

With respect to the water soluble vitamins, insects appear to be closely aligned with known vertebrate nutritional requirements: both groups require about the same vitamins for the same biochemical reactions. This is reasonable because the major energy yielding pathways are similar in both groups. The fat soluble vitamins, vitamins A, D, E, and K, are not involved in the oxidative pathways we have discussed. Vitamins D and K are involved in functions occurring in vertebrates, but not invertebrates. Vitamin D is involved in bone metabolism, and vitamin K is involved in the vertebrate blood clotting cascade. Vitamin K was discovered in 1929 by a Danish scientist, Henrik Dam, who later won a Nobel prize for the work. The K comes from the Danish word koagulation. The long-standing wisdom holds that insects or other invertebrates do not require vitamins D and K. Vitamin A is involved in many mammalian tissues, mostly in ways not completely understood. One well-appreciated action of this nutrient is its role in vision. We will discuss the physiology of vision in a later lecture, in which the role of vitamin A will be addressed in detail. For now we will just make the point that vitamin A is required for insect vision. Vitamin E is also thought to be dietarily essential in at least some insects. Vitamin E is essential for fertility in rats, and apparently in male crickets.

The specific roles of vitamin E are not understood, although its most well known activity is its action as an antioxidant. Vitamin E is thought to interact with certain free radicals, and thereby prevent oxidative damage to cells. Whether vitamin E is required by all insects is difficult to say. We speculate it is required, however, the requirement is very hard to detect because the vitamin is stored in tissues and turns over only very slowly. Now there is one last problem in energy metabolism. Referring to the glycolytic pathway again, please note that at one point NAD picks up a pair of electrons to become $NADH^{+}$ H^{+}. There are two points to be made: One, if the electrons were not subsequently donated from the NADH, cells would eventually become depleted of NAD - and energy metabolism would come to a screeching halt right at this step.

So it is important to recycle this electron carrier to keep energy metabolism rolling along. The second point is that this pair of electrons represents a certain amount of potential energy in the form of additional ATPs if the electron pair could be shuffled over to the electron transport chain. Stated in problematic terms, we have this problem: the outer mitochondrial membrane is not permeable to NAD. There are two pools of NAD in most cells. One is the inner mitochondrial pool and the other is the cytoplasmic pool. Since the electron transport chain is tucked away in the inner membrane of the mitochondria, how is the electron pair, associated with the cytoplasmic pool of NAD, to be delivered to the electron transport chain, or is it delivered at all?

Let us consider the operation of the glycerol-3-phosphate shuttle. On the left is the glycolytic pathway. This is in the cytosol fraction of the cell, where the NAD accepts an electron pair and becomes NADH (plus H^{+}). The electron pair is transferred to dihydroxyacetone phosphate (DHAP). This step regenerates the NAD, thereby allowing the glycolytic pathway to continue its important operations. Reducing the DHAP by addition of electrons converts it from DHAP to glycerol-3-phosphate. This newly formed compound easily migrates across the outer mitochondrial membrane. Again, the membrane is a barrier to NAD flux. Once in the mitochondria, the electron pair is donated to the electron transport chain at the level of FAD, which converts the glycerol-3-phosphate back into DHAP.

DHAP is free to wander back out into the cytosol. Then two points are covered by the glycerol-3-phosphate shuttle: the NAD is regenerated and the potential ATPs are realised. This shuttle is best known from insect flight muscles. There are other pathways to account for moving reducing equivalents from cytosol into mitochondria, one of which is called the Malate Shuttle because malate is formed by oxidation of oxaloacetic acid. The malate moves into the mitochondria, where it is converted to fumarte and then into succinate. This is a more complex shuttle, and we once speculated at a scientific meeting that the glycerol-3-phosphate shuttle is much faster than the malate shuttle, hence more appropriate to very rapidly metabolizing tissues, such as insect flight muscle.

TERRESTRIAL INSECTS

We can regard the current line of talks as a discussion of physiological adaptations to terrestrialization. Perhaps we can try to integrate several areas of physiology along those lines, including information about the integument, the respiratory system and the circulatory system. In our last meeting, we looked at the business of maintaining an optimal osmotic balance between the two major fluid compartments in insects, the intracellular compartment and the extracellular compartment. We saw that the osmotic balance was, in part, achieved by mechanisms that restrict and offset the unavoidable loss of water that must necessarily influence terrestrial organisms. We emphasize the water loss is unavoidable. In particular, we saw that the properties of the cuticle, control of spiracular openings and resorption of water from faeces are important mechanisms to control water loss. There are also a variety of mechanisms to replenish water supplies.

Today we want to continue to develop our appreciation of osmoregulation by observing that organisms must an optimal balance of water and of salt within the body. We develop the problem by considering some specific data. A brief look at the concentrations of various ions in an insect's regular food and in its hemolymph shows a couple of points. First, the compositions of the the food and of the hemolymph are remarkably different in many cases. Second, and more important, the differences are not uniform across the range of individual ions. Here are examples. Larvae of the bot fly *Gastrophilus* are internal parasites of horses, mules and some larger wild animals. In this insect sodium is much higher in the hemolymph than in the food, while potassium is higher in the food than in the hemolymph. A different situation obtains in the stick insect *Dixippus,* which often eats Privet leaves. Privet comes from shrubs in the olive family. Here, sodium, potassium and calcium are much higher in the food while magnesium is much lower.

The insect would simply limit the absorption of unnecessary salts. They would absorb only the required organic food material while unnecessary inorganics were passed though the insect and voided in due course. However, that is not the way hemolymph ion concentrations are regulated. In point of fact, dietary inorganic salts are readily absorbed across the epithelium of the alimentary canal and the ions therefore mix with the general pool of inorganic material in the extracellular compartment.

The physiological problem, then, becomes one of moving ions into and out of the extracellular compartment in such as way as to sustain an optimal osmotic balance. The most general expression is that secretion occurs in the Malpighian tubules and resorption takes place in the rectum. A generalized model of the alimentary canal shows the foregut, midgut and hindgut. Within these familiar major regions are the Malpighian tubules, which most often join the alimentary tract in the midgut-hindgut junction, and the rectum, an enlarged portion of the hindgut featuring specialized epithelial cells. We want to appreciate the various sorts of movements of materials across the different regions of the alimentary tract. This is to be taken rather as a composite, and represents all sorts of movements, some of which do not occur in all insects. In due course we will see that water, inorganic ions, sugars and amino acids are mostly absorbed in the midgut.

With respect to ion concentrations, we want to see that water and ions are secreted into the lumen of the Malpighian tubules. These components are moved into the hindgut by mass flow, and then on to the rectum. Depending upon the individual needs of the insect, water and ions are reabsorbed by cells in the rectum and moved back into the hemolymph. We might get the impression that the same hemolymph constituents are moved in a sort of futile cycle, but we will see that the situation is otherwise. Malpighian tubules are closed-ended tubules that arise from either the posterior part of the midgut or the anterior portion of the hindgut. The precise anatomical arrangement differs among insect groups. The walls of the tubules are one cell layer thick. The cells comprising the tubule walls stand on a basement membrane. The tubules from most insects are surrounded by muscles which move the tubules around in the hemocoel, and the tubules are supplied with numerous tracheae. Malpighian tubules of mosquitoes and some other insect species do not have muscles.

There are distinct histological and functional regions within the overall structures of Malpighian tubules from many, but not all, insects. The tubules have proximal and distal portions, proximal being closer to the attachment with the alimentary canal. The distal portion is functionally the secretory region, and features cells with honeycomb borders. By secretory we mean that water and ions are secreted from the hemolymph into the lumens of the tubules. The proximal region is composed of cells with brush borders; this is the absorptive region of Malpighian tubules. We want to appreciate the cellular structure. The cells attach to the basement. Cells from the distal region feature a honeycomb border formed by cytoplasmic filaments. These filaments are rich in mitochondria, which supply the energy required for active transport processes. The cell membranes on the basal side are envaginated. All of these features are consistent with the secretory function of these distal cells. Cells from the proximal region of the tubule are different from the cells of the distal region in many insects. These cells feature a brush border.

They are made up of more widely separated microvilli and have less complex envaginations of the basal-side membranes. They have more mitochondria in the basal side of the cell, thought to support their reabsorptive role in the tubule. Again, not all insects have Malpighian tubules of this heterogeneity. We also note that some insects, such as aphids, collembola and thrips have no Malpighian tubules or only small absorptive pads called papillae. The Malpighian tubules of some insects have the distal ends adpressed to the sides of the rectum. The ends are surrounded by a membrane known as the perinephric membrane. The perinephric membrane forms a cavity between the rectum and the general hemolymph. This is known as a cryptonephretic arrangement. In the meal worm beetle *Tenebrio* this is thought to help absorb a maximum amount of water from feces.

A primary urine that is isosmotic to the hemolymph is formed in the Malpighian tubules. A general physiological observation is that water is rarely moved across membranes by itself. One way to remember this is to recall that ions can be moved by active pumping processes and water follows the locally formed gradient. In Malpighian tubules, fluid secretion is driven by active transport of potassium and chloride. In some blood feeders sodium is also actively transported. Potassium chloride (KCl) is usually the major solute in the lumenal fluid, and the concentrations of the ions are usually low compared to the hemolymph. The Malpighian tubule fluid makes its way to the alimentary canal by bulk movement aided by muscles. The fluid then moves through the hindgut to the rectum. In the rectum selective reabsorption of water, ions and various metabolites occurs by active transport processes. It is only in the rectum that the urine becomes strongly different in osmotic concentration from the hemolymph. Both tubular secretion and rectal absorption are under separate endocrine controls which together determine the amount and ionic composition of water that is excreted.

This drawing is derived from the blood-sucking bug Rhodnius. This represents those insects with heterogeneous Malpighian tubules. In the distal region of the tubules potassium, chloride and sodium are pumped into the lumen by active processes; water follows the ion gradient into the lumen. We will talk about uric acid formation and excretion later, but for now please note that uric acid also moves into the tubule. Resorption occurs in the proximal portion of the tubule, with potassium moved by active processes, and water following. After movement of urine and uric acid crystals to the rectum, more water can be resorbed, along with selected ions.

Some insects have homogeneous Malpighian tubules, such as the stick insect. Secretion takes place all along the homogenous Malpighian tubules. Some movements are active processes, such as sodium and potassium pumps, while others are passive. There is no resorption until urine is moved into the rectum, where several ions, including amino acids are moved back into the hemolymph by active processes. These data show specific reabsorption of ions on an individual basis. The hemolymph is typically high in sodium, while potassium is the major urine ion in the tubules.

These data indicate that sodium and chloride are selectively resorbed in the rectum. The osmotic concentration of the hemolymph and the urine are about the same, although of differing ionic compositions. The fluid components of the rectal contents are much more concentrated than the other fluids. These data come from locusts fed with fresh water. By way of helping to prepare for any future exams that could crop up right out of a clear blue sky, consider what these data might look like if the locust had been fed salt water. There would be high sodium levels in the rectum, as an avenue to excrete excess sodium. Again, ions are individually regulated and transported. The key point is that the Malpighian tubules secrete water and salts in urine formation. Superimposed upon this urine formation mechanism is a resorption mechanism to facilitate the selective uptake from the urine of those specific things required to sustain the optimal osmotic balance between the intracellular and extracellular fluid compartments.

Consider these situations that an insect may confront:

- *Subjected to greater than normal desiccation*: This situation may result in increased hemolymph osmotic pressure. The insect may respond with decreased urine formation, and greater rectal water uptake.
- *Subjected to greater than normal desiccation and a food higher than normal in salt*: This situation may result in increased hemolymph osmotic pressure, but the increased salt would not allow a maximal decrease in urine production. The insect may respond with a greater uptake of rectal water and a decrease in salt resorption.

From a biological point of view, Malpighian tubules in some insects are adaptive in unusual functions. Larvae of the face fly *Musca autumnalis* have four Malpighian tubules, two anterior and two posterior tubules. Both pairs arise from a single junction between the mid- and hind-guts. The posterior set is found in close association with the midgut. The anterior set expands distally, and is filled with a granulated liquid. The posterior Malpighian tubules are regarded as ordinary secretory organs. The anterior tubules play a special role in the physiology of pupation.

The puparium of this fly is hardened by deposition of calcium salts, rather than by the usual phenolic cross- linking typical of other insects. Dietary calcium salts, as well as other minerals, are stored in the anterior Malpighian tubules during the larval stages. At pupation, the calcium salts become dissolved, then transported from the tubules into the hemolymph. The calcium is directly transported from the hemolymph to the cuticle. The Malpighian tubules of lacewing larvae undergo morphological changes before pupation. The modified tubules then produce silk for the pupal stage.

WATER BALANCE IN TERRESTRIAL INSECTS

We have spent a little bit of time considering aspects of movement. We've gotten an idea of the contractile process in muscle, a process driven by hydrolysis of ATP. We saw that ATP is generated as an end product of energy metabolism and considered gas exchange as a requisite for energy metabolism. Flight is very important to many aspects of insect biology, and we purposefully emphasize flight. Let us also emphasize the broader biological issue in flight. Flight can be taken under the mantle of motility, or movement. In this view, flight is one of many kinds of movement, all fundamental processes in biology.

Let us now turn attention to still another are of physiology confronting nearly every life form: osmotic balance. Osmotic regulation is the business of maintaining optimal salt and water concentrations in an organism. We begin by considering the biological significance of this. About 50 to 90 per cent of the body weight of an insect, or nearly any other animal, is water. Some of this fluid is inside the various cells that make up the body while the remainder is in the spaces outside of the cells. As a unit, these spaces are regarded as a collective known as the extracellular space. The extracellular space is filled with fluid, for obvious reasons called the extracellular fluid. Among the dissolved constituents of the extracellular fluids are the ions and the nutrients needed by the cells for maintenance of life. The extracellular fluid is in constant motion throughout the body, and it is rapidly mixed to near homogeneity by hemolymph circulation and by diffusion. Because of these mixing processes, we can say that all cells in the body live in essentially the same environment. The great French physiologist, Bernard, called the extracellular fluid the melieu int'erieur or what we would call the internal environment.

The point is to recognize two main fluid compartments in an insect body. One is the intracellular compartment and the other is the extracellular compartment. To appropriate a teaching device from Daniel Matzia, the famed cell biologist retired from UC Berkeley, we can regard the extracellular fluids as the *out environment,* that is, the environment inside the body that cells experience. The intracellular environment, then, could be the *in* environment. However, we know eucaryotic cells have organells and compartments within them. Within cells, the compartment inside any particular organelle, such as a nucleus or mitochondrion, could be the *in-in environment.*

At the organismal level, the world we live in is the *out-out environment.* This language allows us to appreciate the world in terms of environments that things experience: the in-in, the in, the out and the out-out environments. Cells experience the out environment, the same thing that Bernard called the *melieu interieur.* Some percentage of the total body fluids of any organism are intracellular fluids. The fluid within each cell is made up of its own individual mixture of various constituents, such as ions, amino acids, proteins, sugars and so forth. On the other hand, within an organism, the concentrations of these things appear to be reasonably similar from cell to cell. For our discussion, the intracellular fluids of all of the different cells of an organism can be considered to be one large fluid compartment: the in environment or the intracellular compartment. Regardless of the particular language or expression that is easiest to keep in mind, we want to come to the point of view that an insect body has two major fluid compartments: the intracellular and extracellular compartments. The extracellular compartment is easy to visualize as the circulating hemolymph.

While the extracellular environment is substantially different in composition from the intracellular environment, the two compartments are in osmotic balance. That is to say that there is rarely a net flow of water from one compartment to another. There are various mechanisms to handle the maintenance of the correct concentrations of the intracellular constituents, including active transport and facilitated diffusion. These mechanisms are a related, but different matter. The thrust for our purposes is that water always tends to diffuse by a process called osmosis across cellular membranes from the side with fewer total dissolved particles to the side with a greater number of dissolved particles. Let us consider some possibilities. First, if a cell in placed in a solution with far fewer osmotic particles than the cell, water will diffuse into the cell. This can result in considerable swelling, even to the point of bursting the cell. Alternatively, if a cell is placed into a solution with a far greater number of particles, water tends to diffuse, once again, down its osmotic gradient, in this case, out of the cell. A badly shrunken cell that may or may not survive may result from this flow. We gather the significance of a physical phenomenon, osmosis. Osmosis explains the importance of maintaining a proper balance of water and of salts in the extracellular compartment or the melieu int'erieur.

During times when, for whatever reason, the extracellular compartment is not correctly regulated, the physiological situation is exactly like placing cells into a solution that is too dilute or too concentrated. This is a common problem faced by insects and most other organisms living in freshwater, salt water or in terrestrial environments. Again, terrestrial insects are faced with the problem of regulating their levels of body fluids and ionic concentrations such that the extracellular and intracellular fluid compartments are maintained in osmotic balance. Failure in this respect would lead to this situation: the body's cells are effectively placed in a fluid that would cause them to swell or shrink to a point of dysfunction or death. Physiological solutions to the problem of water balance have tremendous evolutionary and ecological significance.

First, it can be argued that the spectacular evolutionary radiation of insects occurred after a host of adaptations to terrestrial life. We saw some of these adaptations during our consideration of movement. Physiological regulation of water balance is another adaptation that supports terrestrial life. The biogeographical distribution of insects reflects the broad range of terrestrial habitats in which insects live and thrive. Many of these habitats are extremely arid and very hot. Physiological adaptations to these sorts of environments have permitted exploitation of such habitats that would otherwise be unavailable to insects. The focus on terrestrial life is not meant to overlook aquatic insects, which we will discuss in future. Furthermore, aquatic insects arose from their terrestrial counterparts. Hence, adaptations to aquatic life can be thought of as a component of successful terrestrialization.

Two main factors may work to upset the essential osmotic equilibrium within an insect's body:

1. Loss of water by transpiration across the cuticle, or a very large gain of water by drinking or eating.
2. The intake of food whose concentration of inorganic salts is quite different from the ionic composition of the hemolymph.

Today we are going to begin with the first point, the regulation of water balance, which may be insulted by substantial loss or gain of water. Insects are generally small creatures, which tends to put them in an unfavourable setting with respect to the ratio of body surface area to body volume. The idea is to recognize a relatively large surface across which water can be lost by transpiration. We will see that water loss across the body surface can not be entirely stopped; however, if it were not restricted to a level that can be balanced by water uptake, insects would not survive in the more arid environments. The first line of defence was the development of a cuticle relatively impermeable to water.

The outer layer of an insect is the integument, which is made up of the epidermis and the cuticle. The epidermis is the outer cell layer. There are three major sections of cuticle, endocuticle, exocuticle and epicuticle. In standard drawings, we note the pore canal, cells of the epidermis and the basement membrane. The endocuticle provides the extensibility of the integument and combines toughness with flexibility. The rigidity of the hard parts of an insect, the appendages and so on, comes from the properties of the exocuticle. The relative impermeability of the cuticle to water is due to the presence of the epicuticle. The epicuticle is a complex structure that differs among various insects. There is a cuticulin layer made of two components: a thin, dense outer layer and a thicker, still denser inner layer. Cuticulin is generally made of lipoprotein. A waxy layer lies atop the cuticulin. This is made of long-chain hydrocarbons, and esters of fatty acids and alcohols. There is an inner wax layer composed of strongly oriented long-chain molecules that generally form a monolayer only one molecule thick. The strong orientation provides the waterproof properties of the epicuticle. The outer wax layer is composed of wax molecules that are randomly orientated; on top of this is the cement layer.

People think that this is a shellac-like substance that may serve to protect the wax. Terrestrial insects lose water by evaporation from the general body surface - a process called transpiration. The rate of water loss from the surface on any given insect is a function of both wind speed and also of the humidity of air. Transpiration is relatively independent of body temperature, up to some point. Water is lost from the body surface at a constant, but rather low, rate through a broad range of temperatures. At some temperature the cuticle reaches a temperature point called the transition point.

This point varies from insect to insect, depending upon the chemical composition of the hydrocarbons and waxes in the epicuticle. Insects living in hot, arid habitats tend to have wax layers with higher transition temperatures than do insects living in cooler or more humid habitats. Some insects have two transitions points. Water loss from the body surface is held to a low rate by the oriented wax layer on the outside of the cuticulin. When the temperature of the cuticle rises to such a point that the van der Waals forces holding the wax molecules together are broken, the molecules move slightly apart and water is allowed to escape. Below the transition temperatures the wax molecules are close together, and at a slight angle so that water loss is restricted. At temperatures above the the transition point, the molecules part, which usually leads to a quick and probably pleasing and comfortable death due to dehydration.

Then we see that osmotic balance in the melieu int'erieur is in part maintained by reduction of water loss from the body surface. Another important avenue of water loss in terrestrial insects is from the respiratory surfaces. In the orthopteran

Gastrimargus, for example, 70 per cent of the normal water loss is from the trachael system. We have already seen that water loss from the respiratory surfaces is restricted by the invagination, or internalization, of the trachael system such that it is inside the body. Water loss is also reduced by control of the spiracular valves. Internalization of the respiratory surfaces in a major adaptation to terrestrial life in vertebrates as well as in invertebrates.

We can model an experiment designed to demonstrate the influence of spiracles on water loss. Consider the rate of water loss from *Tenebrio* larvae as a function of time. *Tenebrio* is well adapted to arid conditions, and its water loss rates are typically quite low. If we impose the condition of open spiracles, by holding the animals in an atmosphere of 5 per cent carbon dioxide, water loss would be far higher than what could be easily balanced by an insect living in an arid habitat. Physiological regulation of the spiracles plays an important part in reducing potential water loss.

Some insects live through long periods of quiescence during which they do not feed or drink water. Control of the spiracles in these species is aimed at keeping them closed as much as possible to minimize water loss while facilitating the gas exchange necessary to support metabolic needs. One strategy is called cyclic release of carbon dioxide. This has been well studied in pupae of *Hyalopohora cecropia*, a large insect suitable for study in this regard.

Carbon dioxide is allowed to build up in the trachael system for a long as seven hours, then is released in bursts during which the spiracles are wide open. Between the bursts of carbon dioxide release the spiracles are not closed as tight as possible, but rather flutter so as to allow a continuous uptake of oxygen. While the cuticle and control of the spiracles serve to restrict the loss of water, a certain amount of water is inevitably lost.

Then we can identify three avenues of water loss:

1. Transpiration across the cuticular surfaces
2. Transpiration across respiratory surfaces
3. Loss of water in the excreta

Water that is lost must be balanced by an appropriate level of water uptake. Most insects take in sufficient water with the food they eat. In his book on insect structures and function, Chapman thought some insects may select food with a higher water content under certain circumstances. Still other insect species may consume more food than they actually need in order to realise the potential water in the food. Moreover, many insects drink water. The dipteran *Phormia,* for example, has specific water receptors in the tarsal and labellar sensilla. I've often observed wasps drinking water in desert pools in Mexico. Some insects have evolved slightly more bizarre adaptations to gain water. Larvae of the dipteran, *Epistrophe,* can extrude an anal pipilla into a water droplet and absorb water. The cockroach *Periplaneta* is among many insects that can absorb water from water droplets on the cuticle.

The cuticle is asymmetrical with respect to water movement: water passes inwards more readily than outwards. Also, a few insects can absorb water from water vapour in the air when they are desiccated. *Tenebrio,* for example, can absorb water from air with relative humidities greater than 90 per cent. Prepupa of the flea *Xinopsylla* can absorb water from air at relative humidity above 50 per cent. A final source of water is called metabolic water. As we saw during our look at metabolism, the final electron acceptor at the end of the electron transport chain is oxygen. Electrons and their hydrogens combine with oxygen at this point, yielding water. The amount of water that can be gained by metabolism depends on the metabolic substrate. A hundred grams of the fatty acid palmitate (16:0) yields 112 grams of water while 100 grams of glycogen yields 56 grams of water. Grain insects, such as *Tribolium,* normally obtain a very large proportion if not all of their water from metabolism of their food. We can guess that metabolism of lipid may be adaptive with respect to water balance during long periods of hibernation in several insect species.

8

Mating in Terrestrial Insect

The idea of neuroendocrine systems draws attention to the physiological basis of insect mating systems. A very common example comes from considering the roles and actions of sex pheromones in the mate-seeking phases of mating systems. Lurking behind these components, or phases, of mating systems is still another aspect of insect reproduction. Males have another evolutionary issue. Beyond locating, courting, and mating females, males are constrained by a need to assure the paternity of self. Strong selective pressures seem to favour mechanisms that assure an individual's paternity.

We would just speculate that individuals lacking paternity assurance mechanisms would probably not be well represented in succeeding generations. Insects express many behavioural mechanisms, such as mate guarding, to assure an individual's paternity. Insect also elaborate a number of biochemical and physiological mechanisms to enhance an individual's chances of assuring paternity. Males and females seem to cooperate in some of these mechanisms. The close connection between physiology and behaviour in insect systems. We will consider a mating system because we have a fair bit of experience in this work, and because there sufficient information to present something close to a detailed picture of the system. The mating system of the Austrialian field cricket, *Teleogryllus commodus,* features behavioural and biochemical mechanisms of assuring paternity.

MATING SYSTEM

Nearly everybody knows that male crickets attract their

potential mates by emitting a calling song. Singing is an expensive business for male crickets. Males often maintain and defend a territory that, will later serve as a courtship and mating arena. Male crickets produce their songs by stridulation of the their wings. The singing itself is an energy-intensive process. Before they begin singing, males of some crickets create a small burrow. The burrow acts as a sort of amphitheater, extending the males' broadcast distance. Does the song of a conspecific male sound like some kind of recording star? Maybe a Bob Dylan or a Pavarotti?

We begin with the calling song because the temporal organization of male and female mating behaviours increases the chances of successful mate location for both sexes. First, calling behaviour is under circadian rhythm control. Males express their calling behaviour on a daily basis in a predictable rhythm. Of course, females are attracted to the calling song. Males must be prepared to court and mate with a female by the time he begins his singing.

There are physiological and behavioural preparations. Again, many males prepare a burrow. There is more. As we already know, most (but not all) cricket species mate by transferring a spermatophore from male to female during copulation. Males of *T. commodus* synthesize a new spermatophore each day. As seen for singing, spermatophore synthesis is also under circadian control. In the cricket we worked on, spermatophore synthesis takes place in late afternoon, before nightly singing. The point to take is that two aspects of the mating system are expressed in a regulated temporal organization.

Female crickets also exhibit a number of sexual behaviours. One is her nightly locomotor behaviour. We can show this in an actogram. The top bar shows a 12:12 hour photoperiod. We take three points. First, virgin females express a nightly locomotor behaviour, that is, they go out walking around the park. We might presume they walk around listening for calling songs. The temporal organization of the female walking happens to match up with the male singing rhythm. Hmmm..

The second point is shown in the column under 'S'. The call song can stimulate virgin females into locomotor activity is the song is played during normally inactive periods. We guess this helps out in cases where the clocks are set just right. The third point is that after mating, the female is no longer responsive to calling songs, and she no longer expresses here nightly locomotor activity. The main point here is the sexual berhavior of males and females is characterized by a temporal organization that facilitates mate finding.

Females change their sexual behaviours after mating. Females are no longer stimulated to walk by calling songs, and the nightly locomotor behaviour ceases. We might think these two points make sense because a mated female no longer needs to risk her life walking around in dangerous parks. These behavioural changes also facilitate paternity assurance because the a mated female is more likely to remain in the male's territory. Females also exhibit a third behaviour right after mating. There is a very large increase in egg-laying behaviour, usually in the first 24 hours following mating.

We have a three-dimensional plot of egg-laying behaviour. The x-axis indicates individual females. The y-axis indicates numbers of eggs deposited by each individual, and the z-axis shows four time points. The first x-axis represents egg-laying behaviour in the 24-hour period before mating. This shows that sexually mature virgin females will typically deposit a small number of sterile eggs. The next axis represent the first 24 hours after mating, during which most individuals have deposited a vastly increased number of eggs. The next line shows some females will wait for another 24 hours before laying her eggs. This occurs more rarely than suggested by these data, because the nearly all the females we have experienced will deposit their eggs during the first night.

The burst of egg-laying activity has been called an "egg-flood" because so many eggs are deposited. This is a misleading term because each egg is deposited in a controlled behavioural programme. The eggs do not "flood" out of the females. The main points here relate to females. The temporal organization of female sexual behaviour facilitates mate finding, and mating radically alters female behaviours. These changes in female sexual behaviour, reduced response to male calling songs, reduced locomoter activity and a sustained programme of egg-laying, are cooperative. Together, these changes form the characteristic post-mating behavioural syndrome. The physiological basis of changing the first two behaviours remains unknown. On the other hand, the mechanism of releasing oviposition behaviours is better understood. We know that the release of egg-laying behaviour is not physiologically coupled to the other behaviours.

Let us consider the anatomy of cricket female reproductive tracts. In this drawing one ovary is absent to show structures that otherwise would be obscured. OP indicates the ovipositor, GC the genital chamber, ACG the accessory glands, M the oviducal muscles, LOD the lateral oviduct, LAG the last abdominal ganglion, SPD the spermathecal duct, SP the spermatheca and OV the ovary. A successful mating entails placing a spermatophore into the genital chamber. A fine spermatophore tube makes its way through a pore in the ceiling of the genital chamber, and up into the spermatheca duct. This step is necessary for complete transfer of the seminal fluids into the spermatheca. A fair proportion of matings fail on this point, and examination of spermathecae from these failed matings shows the seminal fluids did not reach the spermatheca.

Aside from sperm, the enzymes and substrate required for biosynthesis of PGE2 are transferred along with the seminal fluids. Once the seminal fluids reach to spermatheca, the transferred enzyme activity converts arachidonic acid into PGE2. The PGE2 moves out from the spermatheca and into general circulation. We believe, with knowing for sure, the circulating PGE2 releases egg-laying behaviour by interacting with specific receptors in the last abdominal ganglion. This is known as the enzyme transfer model of releasing egg-laying behaviour. The model is pretty well supported, although we do not have data on the last point.

Here is a summary of the evidence on the enzyme transfer model:

- PGE2 releases egg-laying behaviour in sexually mature virgin females

- Spermathecae from mated females contain about 500 ng PGE2/spermatheca. Those from virgins had no detectable PGE2
- Spermathecae from mated females synthesized PGE2 at substantial rates, over 30 pmol/hr; those from virgins synthesized none
- Spermatophore contents synthesized PGE2 at about the same rates seen in females, 25 pmol/hr. The spermatophores contained almost zero PGE2

On the first point, when PGE2 was injected into the hemocoel of sexually mature virgins, the females responded by releasing egg-laying behaviour during the following 12 hours. Second, PGE2 could be detected in spermathecae from mated, but not virgin females. Spermatophores themselves contained only very small amounts of PGE2, indicating that PGE2, *per se*, is not transferred to females during mating. third, PGE2 biosynthetic activity was detected in spermathecae from mated females, and in spermatophores, but not in spermathecae from virgin females. Taken together, these data strongly supported the enzyme transfer model.

The next question to ask relates to the specificity of PGE2 in releasing egg-laying behaviour. Do all PGs release the behaviour, or is the action specific to E-type PGs? We understand from our consideration of receptor sites and signal transduction mechanisms this is an important question. If the PGs act in a receptor-mediated physiology, then we would expect to observe specificity to E-type PGs. Let us consider structures of PGs. All 1-series PGs are formed from 20:3n-6. The 2-series PGs are formed from 20:4n-6, and the 3-series PGs are made from 20:5n-3. PGs of each series have 2 fewer double bonds than their parental fatty acids. The key point is most PGs look pretty much alike. They are all 20-carbon carboxylic acids with substitutions on C-15 and a 5-membered ring. The substitutions on the ring define the letter series of each PG.

To investigate the relationship between PG structure and egg-laying activity, we injected a standard dose of PG into the hemocoel of sexually mature virgins, then observed the egg laying responses. We worked with 16 different compounds.

We arranged these structures into three groups, based on structural features. The compounds in group 3 are not the standard PGs. We tested the egg-laying activities of a monounsaturated fatty acid, to control for a general pharmacological effect of lipids. We used a PG that is sterically indered by an oxygen bridge, and a lipoxygenase product that lacks the 5-membered ring of PGs. None of these compounds release egg-laying behaviours.

At the other extreme, highest egg-laying activity was associated with the group 1 PGs, all variations on PGE2. All feature a 5-membered ring, and in every case the ring is substituted with a C-9 keto and a C-11 hydroxyl. These particular ring features are important. We can compare the egg-laying activity of other PGs lacking this arrangement, those in group 2. The group 2 PGs lack the PGE features, and these yielded zero to only low egg-laying activity. Overall, there is a biological specificity for E-series PGs. Again, this finding is consistent with the idea that the PG action is mediated through receptor sites. The receptor sites can discriminate among closely related molecules.

The enzyme transfer model indicates the enzyme activity from males converts arachidonic acid into PGE2 within the spermathecae of newly mated females. The PGE2 subsequently moves into circulation, then releases the behavioural programme of egg-laying. One of the most direct tests of this idea would be to place radioactive arachidonic acid into spermathecae of mated females, then see if radioactive PGs turn up in their hemolymph. The practical problem is it is next to impossible to inject anything into a spermatheca. After trying this a few times, we conceived the idea of letting a male cricket inject radioactive arachidonic acid into spermathecae.

We started this by injecting about 20 mCi of radioactive arachidonic acid into males on the day they emerged as adults. These males started producing a spermathophore each afternoon about 3 to 5 days later, according to their usual development. Here is where it can be quite useful to know males produce their spermatophores on a circadian rhythm. We collected spermatophores each afternoon, then assayed them for radioactivity. The males produced radioactive spermatophores for the next 49 days. So males are able to produce radioactive spermatophores. I wanted to determine the radioactivity was associated with arachidonic acid, and not some uncharacterized metabolic by-product. We collected a bunch of spermatophores, then extracted total lipids from them. After isolating phospholipids on thin layer chromatography, we prepared methyl esters of the fatty acids. We analysed these by radio-high performance liquid chromatography. There was a single peak of radioactivity corresponding exactly with authentic arachidonic acid. So the males somehow packaged radioactive arachidonic acid into the phospholipid fractions of their spermatophores.

This work set the stage for the experiment we wanted to perform. We was interested in seeing the actual sexual transfer of arachidonic acid from male to female. To look at this we allowed my pre-labeled males to court and mate with untreated virgin females. It usually takes about an hour for the contents of a spermatophore to ooze into a spermatheca. One hour after mating with a pre-labeled male, we isolated the spermatheca from each female. The spermathecae from these matings contained about as much radioactivity as a typical radioactive spermathophore. On the other hand, about 24 hours after mating with pre-labeled males, the spermathecae contained no radioactivity. We see the males can transfer radioactive arachidonic acid to females.

The overall data are consistent with transfer of radioactive arachidonic acid from male to female, followed by movement of the radioactivity from the spermatheca into the hemolymph. Of course, the next experiment was to collect hemolymph from females mated with pre-labeled males, and characterize the radioactivity in the hemolymph. Most of the radioactivity in female hemolymph was associated with PGE2 or polar products of PGE2 metabolism. These findings indicate the PGE2 in the hemolymph of mated females is derived from arachidonic acid from males.

Now, this work was done on a single cricket species, *Teleogryllus commodus*. Overall, the work supported the enzyme transfer model, now amended to indicate males transfers and enzyme and substrate for PG biosynthesis in females. We might think this sort of mechanism occurs in many insect species, but it is not so. PG levels are greatly increased in the female reproductive tracts of many insect species.

These include:

- House cricket
- Silk moth
- Cabbage looper
- Migratory locust
- Housefly

The mechanisms of increasing PG levels differ among species. As we might expect, the mechanism in the house cricket and the locust is just like the Australian field cricket mechanism. The moths transfer preformed PGs, *per se*, in their matings. Males of the housefly mobilize and transfer arachidonic acid, but not the enzyme, during their matings. In all cases, PGs increase in female reproductive tracts following mating.

Do these PG increases always release egg-laying? The answer is certainly not. This has been tested in quite a few species, and so far only crickets and the silk moth seem to positively release egg-laying in response to PGs. We conclude that PGs are involved in other, still unidentified, aspects of insect reproductive biology. Let us return to our theme of assuring paternity, by asking what is the biological significance of the cricket mating system?

First, the reproductive success of an individual male is related to how many of his sperm are successful in fertilizing eggs, not to how many females acquire his sperm. By casting the issue in terms of sperm success, rather than mating success, we introduce an area of evolutionary biology known as sperm competition. Mechanisms to avoid or reduce sperm competition are recognized as a powerful force in the evolutionary shaping of mating systems. Two conditions must obtain to allow the occurrence of sperm competition. One, females must be able to mate with more than one male. Two, sperm must remain viable in the body of the female. Sperm competition is not at all restricted to insect biology. There are many examples of the influence of sperm competition in other invertebrates as well as vertebrate mating systems. One of the basic experimental strategies in assessing sperm competition is known as multiple mating experiments.

In these experiments a single female is allowed to sequentially mate with two or more males. Among insects, the general finding is that the first male does not always fare quite so well. In many experiments the last male's sperm fertilizes 75 to 100 per cent of the eggs. Here are two examples of strategies to increase chances of the mating success of the last male. Fruit flies displace sperm from previous mating. Similarly, locusts "flush" sperm from previous mating.

These actions serve to reduce the competition for egg fertilization posed by the sperm populations already in a female's possession. There are many adaptations, behavioural and physiological, that appear to give a competitive edge to one's own sperm, relative to another population of sperm. Some of these adaptations seem to work in concert. The cricket mating system includes a biochemical mechanism to release egg-laying behaviour.

For another component of the same system, males perform their calling, courtship and mating in a pre-constructed arena. Due to her own syndrome of post-mating behaviours, the females tend to remain in their arenas. For still another component of the system, males tend to guard their newly mated females until the eggs are deposited.

SPERM COMPETITION

Plugs

These serve as physically barriers, blocking the entry of additional sperm populations. These are seen in various insects, spiders, snakes and mammals.

Prolonged Copulation

The male himself acts as a sort of plug. This is seen in many insects and in certain vertebrates.

Post-mating Guarding

Seen in many insects, including dragonflies, dungflies and in the cricket system.

Take-over Avoidance

Includes going to places of low risk, and maintaining territories.

Transfer Chemical agents Altering Behaviour

These agents might suppress female receptivity, or release oviposition prior to subsequent mating.

9

Endocrinology of Insect Development

The events that take place in the egg stage of insect life cycles are known as embryonic development. Upon emerging from eggs, juvenile insects embark on an excursion of post-embryonic development that will eventually take them to their adult forms. Post-embryonic development is characterized by dramatic increases in size and by formation of new anatomical structures characteristic of adults. Argued from an ecological point of view, juvenile stages of insects are devoted to eating, growing, and accumulating energy reserves while adult stages are devoted to migrations, mate-finding and reproduction.

This appears to be particularly true among some insects, such as certain species of sphinx moths, which do not feed during adulthood. But as generalizations, these broadly sweeping remarks are not strictly true. Development of reproductive organs begins early in larval life in many insect species, so even if an insect is not actually reproducing, it is preparing for reproduction quite early in life. Many adult insects spend quite a lot of time in feeding behaviours. The value of broad generalizations is that we can appreciate the major thrust of insect life stages. Certainly larval insects are largely given over to eating and growing.

Insects are invested in a more or less rigid exoskeleton, and a generally accepted wisdom is the physical constraints of a least certain of the very rigid parts of exoskeletons inhibit growth beyond set limits. Here is an example of the overall increase in weight during several weeks of development. The figure emphasizes the potential to increase in size during post-embryonic development, in this species from a few milligrams at hatching to nearly 2 grams.

Fully hardened cuticle does not expand, so growth of the hardest parts can only take place during molts. The softer parts of insects can expand. The softer cuticle often occurs in folds that can be expanded, and sometimes the cuticle itself can be stretched. The soft parts of an insect grow rather continuously, while the hard parts grow in steps.

Growth can refer to an increase in cell number or an increase in cell size. The development and expression of cuticle depends on the cells of the epidermis. Epidermal cell numbers increase prior to molting in many insect species. In larvae of many Diptera, growth is due exclusively to increased size of the epidermal cells. Similarly, growth of internal organs can proceed by increased cell size or increased cell number. Both forms of growth occur in the mosquito *Aedes*. Cell numbers in the nervous system and fat body increase, but most other tissues have a constant number of cells. Growth occurs by cell enlargement in these tissues.

Malpighian tubules are separated into primary Malpighian tubules and secondary Malpighian tubules. The primary tubules arise from cells of the proctodeum during embryonic development. The secondary tubules emerge during post-embryonic development. The secondary tubules are reduced to buds during each molt, then develop by cell division, following by cell enlargement. Let us turn to the endocrinology of molting. We will consider five hormones in the regulation of postembryonic development, plus a newly discovered, sixth hormone.

Prothoracicotropic hormone (PTTH) is synthesized in neurosecretory cells of the brain. It is stored and released from the corpora cardiaca. Again, PTTH is released from the neurohemal portion of the CC. PTTH stimulates the prothoracic glands to release ecdysone into the hemolymph. The prothoracic glands have other names. We will see the prothoracic glands called by other names, including ventral glands and ecdysial glands. In higher Diptera it is part of the ring gland. Ecdysone is not the active molting hormone. Various tissues, including the fat body convert ecdysone to 20-hydroxyecdysone, the active form of molting hormone, shown here.

The cells of the epidermis respond to 20-hydroxyecdysone with initiation of the process of molting. Apolysis refers to the detachment of old cuticle from the epidermis, the first step in molting. Molting fluid is actively transported into the space between the old cuticle and the epidermis. Epidermal growth by cell division may occur during this time. The next step involves depositing layer of new cuticle. This is followed by activation of enzymes in the molting fluid that function to digest inner layers of the old cuticle.

The amount of juvenile hormone (JH) released from the corpora allata (CA) determines the form of the new cuticle that is deposited. When JH is present in high concentrations, the new cuticle is larval; when JH is present in low concentrations, the new cuticle is pupal. Adult cuticle is formed in the absence of JH. Sir Vincent Wigglesworth, who was knighted for his research in medically important insects, first suggested the idea of an "inhibitory hormone" in the 1930's. Working with penultimate instars of the blood-sucking bug, *Rodnius*, Wigglesworth showed that if bugs were decapitated 5 days after a blood meal, they molted into the last larval instar. If the bugs were decapitated 3 days after a blood meal, they molted into precocious, but miniature adults. The Harvard biologist Carrol Williams gave JH its name. Williams worked with *Cecropia* moths. The adult moths were parabiotically joined to diapausing pupae. Parabiotically is a relatively rare word for people who work on mammalian systems.

Parabiotic animals are joined in such a way that they share hemolymph circulation. Wigglesworth was famous for his "parabiotic trains", in which many *Rhodnius* nymphs were joined in long trains; molting would progress rearward down the trains. Williams' idea was to try to prolong the life span of the adult moths, which do not feed, by providing nutrients from the pupae. The interesting thing was this: when the pupae were joined to adult females, the pupae would continue normal development and emerge as adults. Alternatively, when the pupae happened to be joined to male adults, they molted into another PUPAL stage.

A series of now famous experiments finally showed that males of some moths, including *Cecropia* moths, accumulate a blood-borne factor in their abdomens that exerts a "status quo" effect on pupal moths. Williams named the factor juvenile hormone because it caused molting into a second pupal stage.

Ecdysis refers to shedding the old cuticle, and is the basis of the name ecdysone for molting hormone. We note that apolysis and ecdysis are separate processes. In some insects shedding of the pupal skin by adults is triggered by another brain hormone called eclosion hormone (EH). EH is synthesized in the median neurosecretory cells of the brain, and stored in the neurohemal portion of the CC.

It is released from the CC under regulation of a circadian rhythm. EH stimulates the abdominal nervous system to play out a "hard-wired" behavioural programme that helps the adults emerge from the pupal case.

The influence of EH on nerve firing in an in vitro preparation of the tobacco hornworm ventral nerve cord. EH was added to the preparation 22 minutes before the first spike occurred. The early spikes (A and B) represent the rotational bursts that begin ecdysis. The later spikes (C and D) represent peristaltic bursts. Jim Truman and Lynn Riddiford, then at Harvard, are responsible for the discovery of EH, and its role in helping insects emerge from the pupal case. Truman and Riddiford married, and took up dual positions in the Department of Zoology at the University of Washington in Seattle.

Let us now introduce a newly discovered hormone in the physiology of insect development. While molting appears to be rather abrupt, the process is long and detailed. As molting progresses, the inner layers of the old cuticle are digested away. Then a soft new cuticle is deposited. When these developmental events are complete, the shedding processes begin, and the new cuticle is expanded and hardened. The key point is noting the shedding processes are behavioural, and the role of EH is to release a pre-programmed sequence of behaviours. We now know the release of the ecdysial behaviours is more complicated. Last year Zitnan and colleagues, then at UC Riverside, discovered the existence of a second ecdysial hormone, called Mas-ETH. Mas comes from *Manduca sexta*, and ETH stands for ecdysis-triggering hormone. ETH is a 26-amino acid peptide. It is produced by a system of epitracheal glands. In *M. sexta* there are 8 pairs of epitracheal glands, each located on the outer surface of the large tracheal tube immediately adjacent to each spiracle.

The hormone is produced and released from Inka cells. Inka is the name of a fairy goddess, known from the Tatra Mountains. Zitnan takes great inspiration from Inka, and he named the cells in her honour. Ecdysis is now thought to follow from the sequential release of two hormones. EH is released from the brain, and it primes the Inka cells to release ETH. The Inka cells release ETH, and the ETH acts on the abdominal ganglia to release the behaviours required for ecdysis. But why do we need two hormones to release ecdysial behaviours? Ecdysis involves many behaviours that are unique to ecdysis. Successful ecdysis requires that all behaviours are coordinated. Failures in coordination result in deformities and death: each ecdysis episode is a crucial event in an insect's life.

The Inka cells are positioned near each spiracle, where tracheal linings are withdrawn at ecdysis. These linings are fragile, and if torn the tracheal supply to that region is blocked. It is thought the Inka cells are in best position to serve as a last inspection site before the insect is irrevocably committed to going through ecdysis.

The physiology of molting involves apolysis, depositing new cuticle and ecdysis. The circulating titre of JH at the time of the molt determines the developmental fate of the insect. Here is an example of the relationship between ecdysone and JH through 4th and 5th larval instars, the pre-pupal wandering stage and pupal stage of the tobacco hornworm. The solid line shows ecdysone titres.

Ecdysone titre spikes a day or so before molting into the 5th instar, then again just before the wandering stage as a prepupa. There is a small increase in ecdysone titre before the pupal molt. Ecdysone titres reach a higher level and remain elevated during the pupal stage. The dotted line represents JH titres. There are high titres of JH in the 4th instar, which are lower in the 5th instar. A small spike of JH is associated with the wandering phase, and no JH is associated with the pupal stage. JH is also active in adult stages of insect life. JH plays an important role of regulation of vitellogenin synthesis in adult female fat bodies. JH acts directly upon follicular epithelia of ovaries to facilitate uptake of yolk proteins. JH is also active in adult males, where it regulates development of reproductive tract accessory glands.

Hormones are transported in circulating hemolymph. Since the hemolymph bathes virtually every cell in the insect body, circulating hormones have the potential to come into contact with all tissues. Again, specific receptor sites form the connection between a circulating hormone and particular target cell. JH and ecdysteroids are not water soluble, and they are associated with specific high-affinity transport proteins. These transport proteins, then, are an important link in the overall physiology of endocrine regulation of biochemical and physiological events. These proteins are also partly responsible for regulating circulating hormone levels. They can act by protecting the hormone from enzyme degradation and by preventing rapid loss of hormone from the hemolymph through excretion.

The titre or levels of circulating hormones are regulated at several points. Biosynthesis, release, transport and degradation of hormones are regulated events. Biosynthesis of ecdysone depends upon PTTH from the brain. JH biosynthesis is partly regulated by nervous signals from the brain to the CA. Ecdysone is degraded to a biologically inactive form, 3-dehydro-alpha-ecdysone, and 20-hydroxyecdysone may similarly be oxidized to a 3-dehydro form.

There are two major pathways that change JH into biologically inactive forms. The minor pathway is hydration of the epoxide functionality to the diol. The most important mechanism is hydrolysis of the JH ester to its respective acid. JH can be broken down by general carboxyesterases, enzymes that occur throughout the insect body and by JH-specific esterases. The JH-carrier protein protects circulating JH from the general esterases, but not from the JH-specific esterase.

The endocrinology of insect reproduction is not without controversy. Mosquitoes elaborate an egg development neurohormone (EDNH) that is produced in the brain. EDNH is released after females take a blood meal. The hormone acts on the pre-vitellogenic ovary, and stimulates release of ecdysone, which is converted to 20-hydroxyecdysone.

The 20-hydroxyecdysone then acts on the fat body, which results in vitellogenin synthesis, and development of the vitellogenic ovary. The interpretation of these data is that in mosquitoes the ecdysteroid produced by the mosquito ovary acts on the fat body to stimulate synthesis of yolk proteins. This, of course, runs counter to the conventional wisdom that JH is released from the CA, then acts on the fat body to stimulate yolk protein synthesis. Among people who are interested in the topic, there are two camps: one group absolutely disagrees with this idea.

Other reproductive hormone actions have been postulated for some insect species. Ken Davey and his co-workers in Canada suggested that there is an internal feedback control of ovarian development in the bug Rodnius. In this insect an antigonadotropin is thought to inhibit the action of JH on ovarian follicles to prevent from developing patency.

This is regarded as a form of direct ovarian inhibition. Indirect ovarian inhibition is thought to occur in several species of Diptera. The general scheme is that the follicle cells produce an oostatic hormone that acts on the neuroendocrine system, causing a reduction of JH.

HEMOCYTIC IMMUNITY

Insects defend themselves from challenges by microorganisms with a combination of humoural and cellular responses. The distinction between these two response categories is rather artificial for a couple of reasons. First, hemocytes can release some of the protein factors that follow from challenges. Hemocytes produce what we regard as humoural responses. Second, although infections elicit a complex of multifaceted responses, the responses are coordinated and integrated into a single theme: host defence.

Nevertheless, it is very convenient to treat hemocytic and humoural immunity as separate topics. One point of separation is that hemocytic responses are rapid. They occur almost immediately an insect is invaded by microorganisms. Humoural responses are much slower, they are usually not seen until some hours after infection. These longer-term responses may be more important in sustaining the hemocytic host defence reactions initiated earlier in the infection cycle.

Circulating cells are regularly found in the hemolymph of insects. These cells are known as hemocytes. Hemocytes are involved in conferring a cellular, or hemocytic, immunity to insects. These cells are formed in haemopoietic organs that are variously well developed in different insect groups. In the cricket *Gryllus* haemopoietic organs are well developed, distinct organs found in the second and third abdominal segments along either side of the heart. In other insects haemopoietic organs are amorphous clumps of cells irregularly distributed along the dorsal vessel.

Plasmatocytes (PLs) are larger cells involved in major immunological actions. PLs are capable of phagocytosis, a kind of entocytosis of smaller foreigh cells. Phagocytosis usually follows low doses of invading yeast and bacteria. PLs are also involved in encapsulation of clumps of invading microbes. There are typically more PLs than other types of hemocytes.

Granulocytes (GRs) are characterized by the presences of many intracellular granules. Electron microscope studies have revealed differences among the granules in these cells. There are structureless and structured granules. The biochemistry of the different types of granules is not yet understood. GRs are major players in encapsulation reactions. Other hemocytes include spherulocytes, oenocytoids and adipohemocytes. These cells are regarded as "nonimmu-nocompetent" by some authors, although there is considerable ground to disagree with this idea. Gupta prefers to use the term "immunocytes" for GRs and PLs and "hemocytes" for all the other types.

He uses these terms to stress the idea that GRs and PLs make up most of the immune responses. Spherulocytes may be involved in silk formation, melanization and secretion of some hemolymph proteins. They may also be important in hemolymph clotting responses to wounds. Oenocytoids may contribute to the final stages of cuticle formation. Other authors suggested that these cells may recognize "non-self".

DEFENCE MECHANISMS

Again, hemocytic defence responses occur immediately insects are invaded by microorganisms. GRs and PLs are responsible for a number of cellular immune reactions. They are able to phagocytize small biologic particles such as bacteria and yeast. Phagocytization refers to the process of cellular internalization of foreign invaders. Phagocytosis is not the same as killing the microbial cells. Microbes are secondarily killed after phagoccytosis.

GRs and oenocytoids are attracted to sites of wounds, where they play major roles in coagulating hemolymph to stop bleedings. We know virtually nothing about the signalling process that is responsible for moving cells to wound sites.

Another defence mechanism is nodule formation. Nodules are aggregates of hemocytes that entrap invading microbes. This is a particularly effective means of clearing the hemolymph, and it is considered to be of greater importance than phagocytosis in clearing large doses of bacteria. There appears to be a sequence of trapping by GRs, followed by aggregation by PLs. The precise sequence of events will remain a point of controversy. In addition to bacteria, viral polyhedrons, fungal spores and protozoans are taken up in nodules.

When organisms or clumps of organisms are too big to be phagocytized or to form nodules, they may be encapsulated. Encapsulation is the sequesterization of these organisms or clumps with multilayered aggregates of hemocytes. This also involves release of coagulum, in a response similar to wound-repair. The coagulum probably aides in forming layers of cells. Encapsulation was first observed by the Russian zoologist Elie Metchnikoff in 1892. He placed rose thorns into a starfish larva.

The next day he saw that the thorns were completely surrounded by tiny cells. Metchnikoff called these cells phagocytes from the Greek words for eating and cells. Metchnikoff eventually moved from the University of Odessa to the Pasteur Institute, and later shared the Nobel Prize for his studies of cellular defences in mammals and invertebrates.

We are not sure of the killing mechanisms that hemocytes elaborate. Some of the suggested mechanisms include peroxidation, melanization, and lysozymes. By comparison to our scant knowledge of invertebrate immunity, very much is understood about the killing mechanisms of mammalian host-defence cells. There are two sorts of mechanisms.

One sort does not depend on oxygen. These include protease, esterase, acid-phosphatase and lysoszyme mechanisms. Another sort does depend on oxygen, including formation of singlet oxygen and superoxide anions, which are very reactive and very toxic species. Work on killing mechanisms of insect hemocytes will produce important and very novel insights into how insects protect themselves.

We recently addressed the question of how the early, hemocytic responses to infection are mediated. What we wanted to know was what biochemical signals occured between recognition of a bacterial infection and the hemocytic responses that follow. Since many aspects of host defence in mammals are mediated prostaglandins and related eicosanoids, we formed the hypothesis that these molecules are also crucial mediators of immunity in insects. The central experimental design was to inhibit various of the eicosanoid biosynthetic pathways, then observe the effects of inhibition on the insects' ability to remove bacterial infections from their hemolymph. Since hemocytic responses occur in the first few minutes post-infection (PI), we could dissect the humoural responses from hemocytic responses by doing the experiments within the first hour PI. In the first experiments, cells of the insect pathogenic bacterium *Serratia marcescens* were injected into 5th instar larvae of the tobacco hornworm *Manduca sexta*.

Test animals were first injected with the phospholipase A2 inhibitor dexamethasone, and control animals were first treated with pure ethanol. Hemolymph samples were withdrawn at 15, 30 45 and 60 minutes PI. The samples were diluted slightly, then streaked onto agar plates. The plates were incubated for about 40 hours, then the number of red colonies on each plate were counted.

The bacteria were used in these experiments are insect pathogens. We reasoned that the dexamethasone decreased the ability of the larvae to clear bacteria from their hemolymph. Since these bacteria are pathogens, the decreased clearance from hemolymph should translate into increased mortality from these infections. We tested this idea by injecting the same 4 doses that were just described into 4 groups of larvae. After 14 hours we counted the number of surviving larvae. As expected, increasing dosages of dexamethasone were associated with decreasing larval survival at 14 hours.

Dexamethasone inhibits the activity of phospholipase A2. This in turn inhibits eicosanoid biosynthesis because substrate is not available to the eicosanoid-biosynthetic pathways. We conducted experiments to test the possibility that arachidonic acid would rescue the larvae from the effects of dexamethasone. These kinds of experiments are called "end product reversal experiments", or something like that. When larvae were treated with dexamethasone, and with arachidonic acid, larval survivorship was a great as survival in the control group.

The phospholipase A2 activity is necessary for all eicosanoid biosynthesis. We wondered if one eicosanoid biosynthetic pathway might be more important than another in mediating insect immune responses. To investigate this point we treated groups of larvae with specific inhibitors of phospholipase A2 (dexamethasone), cyclooxygenase (indomethacin), lipoxygenase (esculetin), and indomethacin and esculetin together. We also observed the effects of maleic acid, an antioxidant compound that does not interfere with the major eicosanoid biosynthesis pathways. After treating the larvae with inhibitors, we infected them with bacteria.

After 2.5 hours, we withdrew hemolymph samples, and plated them on agar plates. We then counted the number of bacterial colonies on each plate. This is a rigorous experiment because if any bacterial colonies were recovered, even a single colony, the individual was counted as positive. These data show that at least 1 colony was recovered from about 20 per cent of the control larvae. Colonies were recovered from about 80 per cent of the phospholipase A2-inhibited larvae.

Bacterial colonies were also recovered from the cyclooxygenase-inhibited and the lipoxygenase-inhibited larvae, but from about 60 per cent, rather than the 80 per cent of phospholipase A2-inhibited larvae. These findings indicated that both pathways were involved in the immune response to infections. This idea was supported by the results from the group that was treated with both inhibitors at the same time: they were not different from the dexamethasone effects.

We concluded that products of cycloxygenase and lipoxygenase pathways were involved in mediating cellular responses to bacterial infections in this insect.

The last comment, of course, opens questions of identifying specific cellular actions that are mediated by eicosanoids. Nodule formation is responsible for clearing the bulk of bacterial infections from circulation. We developed the hypothesis that eicosanoids mediate the nodulation reaction to bacterial infections. Using a line of research similar to the one just described, we found that treating tobacco hornworms with eicosanoid biosynthesis inhibitors just prior to infecting them with bacteria resulted in severe impairment of nodulation. The influence of the inhibitors is expressed in a dose-dependent manner and can be reversed by eicosanoid-precursor polyunsaturated fatty acids.

Again, we concluded that products of cycloxygenase and lipoxygenase pathways were involved in mediating nodulation responses to bacterial infections in this insect.

10

Habitat Selection in Wild Animals

The processes of habitat evaluation are fraught with problems. Some problems are specific to the methods used in the data collection or analyses. Many of these problems have already been recognized, and discussions about them in the literature have prompted a host of evolving techniques. Other problems are inherent in the two most basic assumptions of this approach: that researchers can discern habitat selection or preference from observations of habitat use and that such selection, perceived or real, relates to fitness and hence to population growth rate. My goal is to illuminate the scope of the problems involved in habitat evaluation. Assessments of habitat selection and presumed importance are done so often, and study methods have become so routine, that it is apparent that researchers and managers tend to believe that the major problems have, for the most part, been overcome.

TERMINOLOGY

The word habitat has two distinct usages. The true dictionary definition is the type of place where an animal normally lives or, more specifically, the collection of resources and conditions necessary for its occupancy. Following this definition, habitat is organism specific (*e.g.*, deer habitat, grouse habitat). A second definition is a set of specific environmental features that, for terrestrial animals, is often equated to a plant community, vegetative association, or cover type (*e.g.*, deer use different habitats or habitat types in summer and winter). Nonhabitat could mean either the converse of habitat in the first sense (a setting that an animal does not normally occupy) or the second (a specific vegetative type that the animal views as unsuitable); here, the two meanings of habitat converge. Hall et al. argue that only the first definition of habitat is correct and that the second represents a confusing misuse of the term. They reviewed 50 articles dealing with wildlife—habitat relationships and, based on their definition, found that 82 per cent discussed habitat vaguely or incorrectly. The prevalent use of habitat to mean habitat type, this alternative definition is legitimate and well understood in the wildlife literature.

Habitat selection and preference are also more easily understood in terms of differential use of habitat types. Selection and preference are often used interchangeably in the wildlife literature; however, they have subtly different meanings. The distinction posed by Johnson, who defined selection as the process of choosing resources and preference as the likelihood of a resource being chosen if offered on an equal basis with others. Peek suggested that innate preferences exist even for resources not actually available. Furthering this concept, Rosenzweig and Abramsky characterized preferred habitats as those that confer high fitness and would therefore support a high equilibrium density (in the absence of other confounding factors, such as competitors). Thus use results from selection, selection results from preference, and preference presumably results from resource-specific differential fitness.

In controlled experiments, preferences can be assessed directly by offering equal portions of different resources and observing choices that are made. In the wild, however, preferences must be inferred from patterns of observed use of environments with disparate, patchy, and often varying resources. Generally, the purpose for determining preferences is to evaluate habitat quality or suitability, the ability of the habitat to sustain life and support population growth. Importance of a habitat is its quality relative to other habitats—its contribution to the sustenance of the population. Assessments of habitat quality and importance (*i.e.,* habitat evaluation) are thus based on the presumption that preference, and hence selection, are linked to fitness (reproduction and survival) and that preference can be gleaned from patterns of observed use. Use of habitat is generally considered to be selective if the animal makes choices rather than wandering haphazardly through its environment. Typically, the disproportionate use of a habitat compared to its availability is taken as prima facie evidence of selection. Although technically resource availabilityu encompasses accessibility and procurability, these attributes are difficult to measure, so it seems reasonable to equate habitat availability with abundance (typically measured in terms of area), as is normally done in habitat selection studies.

A habitat that is used more than its availability is considered to be selected for. Conversely, a habitat that is used less than its availability is often referred to as being selected against, or even avoided. This is poor terminology, however, in that it suggests that the animal preferred not to be in that habitat at all, but occasionally just ended up there. Use that is proportional to availability is generally taken to be indicative of lack of selection, which is also unfortunate terminology, as illustrated by the following examples. Consider an animal living in an area with only two habitats and using each in proportion to its availability; from this we might assume that the animal was not exhibiting habitat selection.

However, unless the animal was a very low life form, it certainly made choices as to when it visited each habitat and what it did when it got there; anytime it made a choice, and either stayed or moved, it selected one habitat over the other. Arguably, if one analysed these movements on a short enough time scale, habitat use would be disproportionate to availability, enabling detection of habitat selection. As the time scale is shortened, though, the sheer physical constraint of moving between the two habitats (*i.e.*, the distance between them) also affects their relative use. On the flip side, imagine a dispersing animal attempting to traverse an area with no regard for habitat. If its route was frequently diverted by the presence of other, more dominant resident animals, living in their presumably preferred habitats, the disperser's movements would appear to reflect habitat selection (*i.e.*, selection for habitats not preferred by the residents). Indeed, one could reasonably assert that this represents true habitat selection as defined earlier, in that the disperser chose to avoid habitats with dominant conspecifics and thereby improved its chance of obtaining resources and not getting killed; however, one could also legitimately contend that the disperser was simply exhibiting avoidance of conspecifics, and used whatever cues, including markings, droppings, and possibly habitat characteristics, to do so.

These are not trivial complications, but rather examples of the intrinsic ambiguities associated with the application of these concepts. Terms such as selection and preference can be clearly defined, but not easily measured in the real world.

EVALUATING HABITAT SELECTION

Three general study designs have been used to infer habitat quality. The first, generally called the use—availability design, compares the proportion of time that an animal spends in each available habitat type (generally judged by the number of locations, or less commonly, by the distance traveled) to the relative area of each type. The second, the site attribute design, compares habitat characteristics of sites used by an animal to unused or random sites. These two designs generate measures of selection for various habitats or habitat attributes, and habitat quality or importance is inferred from the magnitude of this apparent selection.

The third method, which the demographic response design, uses a more direct approach for assessing habitat quality by comparing the demographics (density, reproduction, or survival) of animals living in different habitats. This design thus circumvents the need to interpret animal behaviour (habitat choices). Among studies of birds and mammals, the use availability design is the most popular. Habitat-related papers dealing with birds and mammals published in the Journal of Wildlife Management during 1985–1995 and found that most relied on a use—availability study design to assess habitat selection, preference, or quality.

Thomas and Taylor further categorized use—availability studies into three approaches: one in which habitat-use data are collected on animals that are not individually recognizable (*e.g.*, visual sightings or sign), one in which data are collected on individuals (*e.g.*, radiocollared animals) but habitat availability is considered the same for all individuals (so individuals are typically pooled for analysis), and one in which use and availability are measured and compared for each individual. They also reviewed papers published in the Journal of Wildlife Management and found that nearly twice as many studies collected data on individuals but pooled them for analysis than either of the other two approaches.

Studies that pooled animals for analysis have commonly compared frequencies of use and availability for an array of habitats using a chi-square test. Determination of which habitat types were used more or less than expected is generally made by comparing availability of each habitat type to Bonferroni confidence intervals around the percentage use of each type. This procedure was described initially by Neu et al. and clarified by Byers *et al.*, although a more accurate method of constructing such confidence intervals was recently proposed by Cherry. If the areas of available habitats are estimated (*e.g.*, from sampling) rather than measured (*e.g.*, from a map), use and availability should be compared with the chi-square test for homogeneity rather than goodness-of-fit. A chi-square goodness-of-fit test assumes that the availabilities are known constants against which use is compared, so if availabilities are actually estimated, with some sampling error, this test is more prone to indicate selection when there is none.

Various other methods of comparing use and availability have been advanced but less often used in wildlife habitat studies. Ivlev proposed an electivity index to measure relative selection of food items on a scale from "1 to 1; this has since been adopted for some habitat selection studies. However, Chesson noted that Ivlev's index may yield misleading results because it varies with availability even if preference is unchanged, and advocated use of a 0 to 1 index originally proposed by Manly *et al.*, also for feeding preference studies. This Manly—Chesson index is simply the proportional use divided by the proportional availability of each habitat, standardized so the values for all habitats sum to 1. As adapted to habitat studies, it is interpretable as the relative expected use of a habitat had all types been equally available.

Thus in an area with four habitats, an index of 0.25 for each habitat would indicate no preference, whereas deviations from this would indicate relative preference for or against certain habitat types. Heisey and Manly et al. extended this method to test for differences in habitat preference among individuals or sex—age groups, and also showed how to test for statistically significant differences among preferences for different habitat types. Kincaid and Bryant and Kincaid et al. offered an alternative method that scores relative differences between use and availability for habitats defined as geometric vectors. Most studies using these tests pooled data among individuals, so that animal captures, sightings, radiolocations, and so on represented the sample units. Aebischer et al. pointed out that this constitutes pseudoreplication and advised comparing use to availability for each animal individually (*i.e.*, so individuals are the sample units). Several methods have been developed specifically to do this. Of these, the most commonly used is Johnson's, which is based on the difference between the rankings of habitat use and the rankings of habitat availability.

This method also provides a means of detecting statistically significant differences among habitats, not just a relative ordering of their selection. Moreover, because comparisons are made on an individual-animal basis, habitat availability can be considered either within each individual home range, or within the study area as a whole. Johnson defined first-order selection as that which distinguishes the geographic distribution of a species, second-order selection as that which determines the composition of home ranges within a landscape, and thirdorder selection as the relative use of habitats within a home range. Thus, both second-order and third-order selection can be addressed with Johnson's technique; with chi-square tests it is possible but more difficult (because of sample size constraints) to consider both of these levels of selection. Alldredge and Ratti compared four methods (including the chi-square, Johnson's, and two others based on individual-animal comparisons) in simulated conditions and found that none performed consistently better than the others. However, some methods are better suited for given situations. For example, because data for all animals are generally pooled for chi-square tests, unequal sampling among individuals could strongly affect the results if all individuals did not make similar selections.

Conversely, the methods that weight animals equally, regardless of the amount of data collected on each, may be subject to spurious results caused by small sample sizes and variability among individuals. McClean et al. used real data on young turkeys (Meleagris gallopavo), which have fairly narrow and well-known habitat requirements, to compare results of six analytical techniques for assessing habitat selection. In this case, the methods that treat individuals as sample units tended to be less apt to detect habitat selection. Aebischer et al. offered what appears to be an improved procedure for comparing use with availability on an individual animal basis. This method (compositional analysis) has become increasingly popular because it enables assessment of both second-order and third-order selection and yields statistical comparisons (rankings) among habitats.

Additionally, because the data are arranged analogous to an ANOVA, in which between-group differences can be tested against within-group variation among individuals, it provides a means of testing for differences among study sites (*e.g.*, with different habitats, different animal density, or different predators or competitors), seasons or years (*e.g.*, with different food conditions), sex—age groups, or groups of animals with different reproductive outputs or different fates.

SITE ATTRIBUTE DESIGN

Site attribute studies differ from use—availability studies in that they measure a multitude of habitat-related variables at specific sites and attempt to identify the variables and the values of those variables that best characterize sites that are used (often for a specific activity). With this design, the dependent variable is not the amount of use (as with use—availability studies) but simply whether each site was used or unused (or a random location with unknown use); the independent variables can be many and varied. Use—availability studies generally just deal with broad habitat types, or if more variables are considered, they are analysed individually.

A site attribute design was used in 45 (29 per cent) of the habitat selection studies. Of these, 28 were on birds and 17 on mammals. This design requires measurement of habitat variables at some defined site, usually one that serves some biological importance to the animal.

Nest sites of birds are easily defined and biologically important, and hence are often the subject of studies of this nature. Habitat characteristics of breeding territories, drumming sites, and roosting sites also have been investigated. Among mammals, studies have focused on characteristics of feeding sites, food storage sites, resting sites, shelters, wintering areas, or areas recolonized by an expanding population.

Other studies have compared habitat characteristics of randomly located sites to sites where birds or mammals were observed, radiolocated, or known to have been from remaining sign. The statistical procedures used in such studies vary. Most have used multivariate analyses to differentiate combinations of variables that tend to be associated with the used sites. Discriminant function analysis (DFA) is the most popular of these. Logistic regression is an alternative, and is especially useful when the data consist of both discrete and continuous variables or are related to site occupancy in a nonlinear fashion.

DEMOGRAPHIC RESPONSE DESIGN

Ideally, studies should identify relationships between habitat characteristics and the animal's fitness. Studies employing use—availability and site attribute designs assume that certain habitat features are selected because they improve fitness. Demographic response designs attempt to test this more directly. However, refer to the measured demographic parameters in these studies as response variables, they really only represent correlates with given habitats. Most of these investigated differences in animal density among habitats. Fourteen studies, all on birds, related reproduction (*i.e.*, nesting success) to habitat of nest sites. Three studies, two on birds and one on mammals, attempted to find an association between habitat and survival, but only one detected such a relationship.

DEFINING HABITATS

The first prerequisite for assessing habitat selection is that habitats be defined as discrete entities. For use—availability studies in particular, the defined number of habitats can directly affect the results. Yet habitat distinctions often are not clear-cut. A researcher might distinguish two general forest types, uplands and lowlands, or might classify habitats by dominant overstory, or might divide these further by stand age or understory, and so on.

As more types are defined, sample sizes are reduced for observed use of each type, thereby diminishing the power of the statistical tests to distinguish differences between use and availability. Also, because the proportional use and availability of all habitats each sum to 1, the number of habitats distinguished affects all of these proportions.

Aebischer et al. observed that this unit—sum constraint renders invalid many of the statistical tests often employed to compare use and availability because the proportions are not independent. That is, if one habitat type. Has a low proportional use, others will have a correspondingly high use, and if there are only a few types, then the infrequent use of one type will lead to the apparent selection for another. Aebischer et al.'s method of compositional analysis was developed specifically to circumvent this problem.

Not just the number of types, but the criteria used to partition types may greatly affect results. Knight and Morris were able to visually differentiate 13 habitat types on landscape photographs of their study area, but postulated that only two broad classifications were distinguished by red-backed voles (Clethrionomys gapperi), the subject of their study.

After analysis of their data, however, it became clear that from the voles' perspective, at least three functional habitats existed. Another problem is the scale at which habitats are viewed. For example, an animal might appear to select for a certain habitat type, defined by a dominant cover type, whereas in reality it selected for certain specific kinds of sites that just happened to occur more commonly in that cover type than in others. An animal's choice of habitat type is often called macrohabitat selection and the choice of specific sites or patches within habitats is called microhabitat selection.

These may be perfectly hierarchical in that the most preferred microhabitats always occur within the same macrohabitat, in which case an animal may really select initially at the scale of macrohabitat, and then focus on specific sites within it. Schaefer and Messier observed this sort of nested hierarchy across a range of scales for foraging muskoxen (Ovibos moschatus) in the Canadian High Arctic. Alternatively, the distribution of preferred microhabitats could be largely unrelated to the broader habitats defined by the biologist; in this case, a site attribute study might identify characteristics related to preferred microhabitats, whereas a use—availability study would detect no selection at the level of habitat type.

This situation was apparently the case for wood mice (Apodemus sylvaticus) inhabiting arable lands in Great Britain: The mice seemed not to select (based on a use—availability study) from among three types of croplands (macrohabitats), but within each of these croplands they chose microhabitats with a high abundance of certain plants.

In sum, significant challenges in defining habitats include: partitioning them in terms of the features that the animals are selecting for, which are not necessarily the ones we most easily discern; delineating sufficient habitat categories to ensure that the truly important types are not lumped with and thus diluted by less important types; and not diminishing the power to discern selection by parceling out too many types.

MEASURING HABITAT USE

Sample bias is an obvious potential problem in measuring habitat use. Interpretations of habitat use from visual observations of animals or their sign can vary among observers and sightability can vary among types of habitats, both of which can introduce biases in the data. For example, Powell noted that fisher (Martes pennanti) tracks in snow were difficult to follow in habitats with dense vegetation, especially where fishers followed trails of snowshoe hares (Lepus americanus); in this case the bias against observing tracks in dense vegetation merely detracted from the overall conclusion that densely vegetated habitats were frequently used.

Counts of pellet groups (*e.g.*, from ungulates or lagomorphs) may poorly reflect habitat use because defecation rates often vary with the food source, and hence the habitat type. Capture locations may be a poor indicator of habitat use because baits and other trap odors (*e.g.*, from captures of other animals) may affect behaviors in an unpredictable way. Telemetry also may yield biased data on habitat use because the detection of an animal's radio signal may depend on the habitat it is in, and location data obtained by triangulation have inherent associated errors.

Intuitively, and as shown in computer simulations by White and Garrott errors in determining habitat use increase with increased habitat complexity and decreased precision in the telemetry system. Errors do not necessarily introduce bias, but can if patch size differs among habitats (detected use would be underrepresented in habitat types that tend to occur as small patches) or if the animal preferentially used the edge of some habitat types but not others.

Powell reported different perceptions of habitat use of fishers between his study, where he followed tracks in the snow, and another nearby radiotelemetry study; he attributed the difference to error in the telemetry system and consequent incorrect habitat categorization for animals near edges. Nams showed that simply discarding locations because of large telemetry error, as is common practice, exacerbates bias; he offered a procedure for circumventing it, but few studies have used it. Kufeld et al. suggested using the habitat composition of error polygons formed by the triangulation of radio bearings, but this would not alleviate bias. Chapin et al. solved the problem a different way. In a study of habitat use of American martens (Martes americanus), which have a documented affinity for mid- to late-successional forests, they classified telemetry locations that were outside small patches of forest but within telemetry error of the edge as representing use of those forest patches.

Even if habitat use can be measured accurately, biases may result from sampling or analytical procedures. Habitat use may vary by individual, sex—age group, social status, time of day, season, and year, yet many (most) studies pool individuals and do not sample adequately. Schooley reviewed habitat studies published in the Journal of Wildlife Management and found that most lasted only 2 years, and most pooled results across years without testing for annual variation. He used results of a black bear (Ursus americanus) study to show that habitat use can vary annually, and that the data pooled across years can yield misleading results. Beyer and Haufler found that most published studies that they reviewed collected data only during daylight hours; in their study of elk (Cervus elaphus), habitat use differed between day and night.

Arthur and Schwartz reported diurnal and nocturnal differences in habitat use for brown bears (Ursus arctos) that fed at a salmon stream that was used by people during the day; this difference was detected with data from GPS collars, but was not apparent from conventional diurnal telemetry data. Ostfeld et al. and Belk et al. observed sex-related differences in habitats used by ground-dwelling rodents; Belk et al. remarked that combining the two sexes would produce a false perception of habitat use. Paragi et al. observed differences in habitat use of resident and transient martens. Boitani et al. and Macdonald and Courtenay observed individual differences in habitat use, apparently related to social status. Bowers found that habitat use of eastern chipmunks (Tamias striatus) varied significantly with distance from their burrows, a finding noticeable only by considering the data on an individual basis. "It is time," Bowers commented, "that ecologists recognize that microhabitat selection and usage is a process involving individuals, not species." Pooling individuals is common because sample sizes are typically too small to test for selection by individual. However, the statistical tests usually used assume independence among sample units, which is often not the case in studies that consider each location a sample. Some techniques consider animals as sample units, so lack of independence among locations within individuals is not problematic.

However, these methods are still subject to difficulties with lack of independence if animals are gregarious or territorial (social exclusion precludes use of certain habitats), or if the study subjects are related (habitat preferences possibly affected by a common learning experience) or are from the same social group. In an effort to alleviate the problem of a lack of independence among individuals, Neu et al. used groups of moose (Alces alces) and Schaefer and Messier used herds of muskoxen as their sample units, rather than individual animals. Similarly, although Gionfriddo and Krausman monitored habitat use of individual radiocollared mountain sheep (Ovis canadensis), they considered groups of sheep their sample unit.

However, Millspaugh et al. contend that animals in a herd should be considered independent individuals if they congregate because of a resource rather than because of a biological dependence on each other. They provide a hypothetical example with elk, where 99 of 100 radiotagged animals congregated at a winter feeding area in one habitat and the remaining individual used a second habitat; at other times of the year the elk did not associate with each other. In this case, they argue that each radiotagged individual should be considered an independent sample. In contrast, predators that hunt together in a pack and are thus dependent on one another cannot be considered to use habitats independently. Millspaugh et al. recommend tests to evaluate independence of habitat use by seemingly associated individuals.

MEASURING HABITAT AVAILABILITY

Measuring habitat availability is often more problematic than measuring use. Use—availability studies inherently assume that study animals have free and equal access to all habitats considered to be available. That is, at any given moment each study animal should be able to use any available habitat. This assumption may hold if use and availability are measured for each animal individually. However, the assumption may be violated when animals are pooled for analysis and the available habitat is considered to be the same for all, yet some individuals may not even have all habitat types within their home range.

Johnson suggested that the habitat composition of home ranges compared with the habitat composition of some broader area should indicate the level of selection animals exercise when establishing their home range. Often the broader available area is considered to be that encompassed by the composite of the home ranges of all study animals. However, there are several problems with this. First, animals cannot really select their ideal mix of habitats to compose their home range. Animals can only choose home range borders that encompass the best mix of habitats from what exists on the landscape; they cannot alter the mix to suit their needs. By analogy, a person may pick a town to live in, among several available, based on the resources available.

One may also choose where to live within the town, but one cannot alter the layout of the town or the array of features available. Second, animals may not have free and equal access to all areas when establishing their home range. Home ranges may be established near the natal area just because of familiarity with resources or neighboring animals, not any choice related to habitat composition. Analogously, people might remain in their home state or country not because they consciously chose it among all others, but because they never had the opportunity to visit other places, or because moving elsewhere, even if it seemed desirable in some respects, had too many costs. Social constraints also may dictate choice of a home range by precluding access to certain areas. Extending the analogy with people, consider a house to be like a home range and a neighbourhood a composite home range. The first few residents of a neighbourhood might have selected where to live among houses that differed in various ways; however, as more people moved in, the choices narrowed, until no choice remained for the last resident.

If all houses were used, regardless of their quality, one could not discern after the fact which houses were preferred unless the "colonization" process was observed. Fretwell and Lucas proposed a corresponding model for animal populations. In an expanding population, preferred habitats are settled first, but as these are taken, animals are forced to settle in poorer and poorer areas. However, unless they are strictly territorial, their ranges can overlap, so unlike the human example, they can choose to live in a preferred area even though another animal is already there. As animal density increases in the most preferred habitat, however, resources become less available to each individual, so the quality of the habitat from each resident's perspective diminishes.

Thus unless individuals benefit from the presence of others, their home range selection is negatively influenced by conspecific density. Other competing species have a further interacting effect on habitat availability and hence selection. Because competition changes each animal's perception of habitat availability, human measurement of availability, based on the assumption of free and equal access, is inevitably inaccurate. Animals tend to be more uniformly distributed across patchy landscapes than predicted from studies of habitat selection. Another major problem in measuring habitat availability is the recognition and treatment of areas of nonhabitat that may exist within home ranges. Part of the difficulty arises simply because our concept of home range is too nebulous. Home range is typically defined as the area used by an animal for its normal activities, but home range area is a human perception, not a biological entity. Humans may perceive the landscape as a mosaic of habitats that fit together like a jigsaw puzzle, on which are superimposed home ranges of animals. In contrast, animals may perceive the landscape as series of corridors or islands sprinkled in an ocean of nonhabitat. If we unwittingly define available habitat from our human perspective, and include large patches of nonhabitat that the animal does not really perceive as among its choices of places to live, a comparison of use to availability might demonstrate nothing more than avoidance of the nonhabitat.

This would be grossly accurate, but not particularly insightful. An example was presented by Johnson, where mallards (Anas platyrhynchos) rarely used open water areas far from shore, but the area of open water was large. Standard means of comparing use to availability, such as the chi-square test, might show open water to be avoided and all other habitats selected; however, a knowledgeable duck biologist would recognize this as a trivial result, and might elect to exclude this obvious nonhabitat from the analysis. Other cases may not be so clear-cut. Manly et al. presented an example with California quail (Callipepla californica), taken from a study by Stinnett and Klebenow. Bonferroni confidence limits, and hence perceptions of selection, depended on whether habitats that were not used as escape cover when the birds paired for mating were included or excluded from the analysis. In this case the habitats that were not used as escape cover during mating were not obvious nonhabitats because the birds used them in other circumstances and for other activities. In this example, habitats 1, 3, and 4 were used and thus also available.

Habitat 2 (depicted as a swamp) appears to have been traversed, possibly just to get from habitat 1 to habitat 3; if it was used simply because of its location, not because of its habitat-related attributes, a question arises as to whether it should be considered in the analysis. Conversely, although habitat 5 was not used, it may or may not be considered available. Judged within the context of the home range boundaries, point A in habitat 5 appears to be unavailable, yet this point is closer to known locations of the animal than points B or C, which are both within the apparent home range. Habitat availability is a nebulous concept, and thus may be difficult to measure. Similarly, although the figure depicts a travel route, from which relative use of habitats might be deduced, most analyses deal with relative time, not distance, in each habitat (partly because telemetry data are generally comprised of point locations); it is unclear which is really a better measure of use. An advantage of Johnson's technique is that the results are rather robust to inclusion or exclusion of habitat types that are rarely used. A problem with Johnson's technique is that because it is based on rankings of use and availability, habitats will not appear to be selected if their proportionate use is ranked the same as their availability.

Thus even if the animal spends an inordinate amount of time in the habitat that is most available, selection for this habitat will not be detected using this technique because both use and availability are ranked the same. The Manly—Chesson index of habitat selection also does not fluctuate with inclusion or exclusion of seldom-used habitats, and Manly et al. showed that this index is much more versatile than Johnson's in many other respects. Recently, it was adapted by Arthur et al. to handle situations in which habitat availability changes. These authors recognized that habitats available to polar bears (Ursus maritimus) varied with changes in ice conditions and with movements of bears across their enormous home ranges.

Thus they defined availability separately for each radiolocation, using the habitat composition of a circle with a radius (from the radiolocation) equal to the expected distance a bear would travel during the time between radiolocations; habitat availability within these circles was then compared with the type of habitat the bear actually used the next time it was located. Another attribute of Manly et al.'s procedure is that it can be used to analyse data from site attribute studies as well as use—availability studies, although site attribute studies also face problems in assessing availability. If used sites are compared to random sites, the universe from which the random sites are drawn must be defined. As discussed earlier, that universe can be some arbitrarily defined study area, a composite home range of study animals, or each individual home range. Additional difficulties may arise if the comparison is between used and unused sites because errors may arise in distinguishing unused sites (*i.e.*, nonobservation of use may not mean nonuse). Furthermore, unused sites may be vacant for a variety of reasons, some of which are unrelated to the physical habitat (*e.g.*, human disturbance, exploitation, predation, parasites, interspecific competition). Some predictive models have fared poorly when they did not consider such variables.

Geffen *et al.* found, unexpectedly, that Blanford's foxes (Vulpes cana) in desert environments were rarely observed near springs, where water and food were most abundant, probably because this habitat was favoured by and provided cover for potential predators. In order to assess the criteria used by a species in selecting sites, investigators ideally should choose for comparison sites with both available resources and predators (or other confounding agents) present, as well as sites with only one or the other; however, such comparisons are unavailable in most field studies.

If a species is very selective in its choice of sites, differences between used and unused sites may be quite subtle; these subtleties would not be discernible in site attribute studies if the investigator chose unused or random sites that were very different from the used sites. The scale of comparison in this case would be too coarse. In an attempt to circumvent this difficulty, Capen et al. eliminated available sites in habitat types that were "radically different" from those that were used (analogous to eliminating nonhabitats in use—availability studies). Conversely, if in attempting to use a finer scale of comparison one picked random sites from too narrow a universe, such that they were all very similar to the used sites, habitat differences might not be detected if a large portion of the random sites were used.

This points out the advantage of distinguishing unused sites instead of just random sites and of selecting unused sites that are similar in many respects to the used sites. Use—availability studies do not distinguish unused areas and so may be especially prone to problems of too fine or too coarse a scale of comparison. The coarse-scale problem (used and available areas are too dissimilar to detect the true basis for selection) may occur when composition of home ranges or habitat use within home ranges is compared to some broader study area. The fine-scale problem may occur when habitat use is compared to availability within home ranges.

Thus these two scales of comparison may yield different results. McClean et al. examined the effects of varying the definition of available habitat, from the entire study area to progressively smaller-sized circles around individual radiolocations. They found that selection became increasingly difficult to detect as availability was defined by a smaller and smaller area. This result is not surprising because the radiolocation represents use, so habitat composition within smaller areas around each location more closely matches areas of actual use.

ASSESSING HABITAT SELECTION

Perceived habitat selection may vary with the technique chosen to compare use and availability or to compare attributes of used and unused (or available) sites. Some of this variation in perceived selection stems from the fact that different methods actually test different biological hypotheses and some is from the different assumptions inherent in these techniques and their sensitivity to violation of these assumptions. Manly et al.'s technique can handle both use—availability and site attribute study designs. Moreover, it can be performed on an individual animal basis or with pooled data, it can be used to compare habitat selection among groups (*e.g.*, species, sex—age classes, seasons, times of day, times within seasons), and it can incorporate both discrete and continuous variables. For these reasons, it has been heralded as a unified approach.

Manly et al.'s approach generates a resource selection probability function, giving the probability of a site being used as a function of various habitat variables. Each habitat variable can be tested to determine whether it contributes significantly to the probability of use. In the special case of only a single categorical habitat variable (*i.e.*, habitat type), the function reduces to the Manly. An advantage of this index, as discussed earlier, is that it is rather unaffected by the inclusion or exclusion of seldom-used habitats. In this sense, Chesson suggested that the index is a measure of preference that "does not change with density unless behaviour changes" and that it represents the expected use of the various resources if all were equally abundant.

A fatal flaw of habitat selection studies in general, especially use—availability studies, is that they are based on the assumption that the more available a resource is, the more likely an animal should be to use it. This may not be true at all, may be true for only some resources, or may hold only within a narrow range of availabilities. Manly et al. made the explicit assumption, applicable for all models that availability remains constant for the period of study (if availability changes seasonally, data can be analysed by season). This may seem like a benign assumption, but in reality it masks a fundamental weakness of the process. Of what value are measures of selection if they are specific to a single array of habitats?

Measures of selection are supposed to be reflections of inherent preference—expected choices when availabilities of all habitat types are equal—so if selection appears to change as availability changes, then preference cannot be inferred from perceived selection when availabilities of habitats are unequal. In other words, if the goal is to assess habitat preferences for a population of animals based on habitat selection observed among a collection of individuals in that population, then something is amiss if selectivity appears to differ among these individuals simply because they have different habitat compositions available to them.

Consider a human analogy that demonstrates the effects of changes in availability on perceived selection. While at home a person spends 50 per cent of the time sleeping and 20 per cent preparing food and eating meals in the kitchen; the bedroom occupies 20 per cent of the area of the house, and the kitchen 10 per cent. Manly—Chesson selection indices for these rooms would be 0.51 and 0.40, respectively. Now suppose the person feels cramped in the kitchen and moves a wall, making it twice as big, at the expense of a room other than the bedroom. Afterwards the kitchen makes up 20 per cent of the area of the house, the same as the bedroom, but use of the kitchen does not increase because it still takes the same amount of time to prepare and consume meals there. The selection index for the kitchen thus drops to 0.25, despite the fact that it is now more comfortable and better serves its purpose. Moreover, although no changes were made to the bedroom, its selection index improved to 0.63 as a result of the renovations to the kitchen. Superficially, it would appear that the expense for remodeling was not worth it.

Analogously, one might imagine a situation in which an animal used a habitat substantially more than its availability, but used it only for sleeping. If that habitat became more available, the animal would not be expected to sleep more, so its selection for it would appear to decline. A management agency that produced more of this habitat because results of a habitat selection study showed it to be used disproportionate to its availability would be disappointed to find that these efforts made the animal's selection for it drop. These examples demonstrate cases in which the activity requires a fixed amount of time, so increasing availability of the preferred setting for that activity has no effect on how much time is spent there. This situation is just a special case demonstrating the point that use and availability are not inexorably linked.

In the example of the house, the renovated kitchen might entice the person to spend more time there, but only up to a point (one certainly would not sleep there). Conversely, if the dining room had been remodeled at the expense of room in the kitchen, the person might not eat in the kitchen anymore, but, no matter how small it was, still prepare food there. Each room might thus have its own functional relationship between area and use. Similarly, if an animal prefers a certain habitat for resting because it offers protection from predators, it might spend more time resting in a larger patch of that habitat because it offers greater safety than a small patch. Enlarging a patch that offers virtually no predator protection to a size yielding some predator protection might thus cause significantly increased use of the patch; however, additional enlargements might have progressively lesser effects on use because they do not add much predator protection, and eventually further enlargements do nothing, or might even attract a different predator, thus deterring use.

Various scenarios and corresponding relationships between patch size and use are plausible. Considering that the relationship between patch size and use probably varies among habitat types and the mathematical relationship between use and availability also differs among the various selection indices, it seems doubtful that one could assess selection just by comparing relative use to the relative area of different habitats. Mysterud and Ims proposed a logistic regression model to compare use:availability ratios among study subjects that had differing habitat compositions available to them. This model thus provides a test of the assumption that use increases with increased habitat availability.

Their method is applicable to cases in which habitats can be categorized into two discrete types (*e.g.*, forested vs. nonforested, oak vs. nonoak). They reexamined two data sets that Aebischer et al. had analysed using compositional analysis. In one, use increased with increased availability of a habitat for 9 of 12 ring-necked pheasants (Phasianus colchicus); however, three individuals did not fit this trend. In the second example, gray squirrels (Sciurus carolinensis) showed an inverse relationship between use and availability of open habitats within their home ranges (the same unexpected relationship therefore existed for the alternate, forested habitat).

It was surmised that size and interspersion of habitat patches greatly affected the choices that these animals made, more so than just total habitat area. Similarly, Mysterud and Ostbye found that although roe deer (Capreolus capreolus) in winter chose open canopy habitat for feeding and dense canopy for resting, they had to balance the advantages of being in each type of habitat against the energetic disadvantages of traveling between them, so patch size (distance between patches) affected habitat selection.

Mysterud et al. suggested that for animals such as roe deer, which face tradeoffs in using different habitats, selection is not directly related to resource availability, so habitat rankings based simply on ratios of use to availability often are misleading. Bowyer et al. used a site attribute analysis to examine habitat selection related to various tradeoffs faced by black-tailed deer (Odocoileus hemionus). Assessing selection can be extraordinarily complex because each habitat is not a single patch, but a series of patches of different sizes and shapes, each bordering other patches of different sizes, shapes, and habitat types. Otis offered a model that tests for the disproportionate use of habitat types as well as habitat patches, thereby providing a means of assessing things such as minimum patch size requirements. Data for this model (patch size distributions for each habitat type and locations of animals in specific patches) are available with modern geographic information system (GIS) coverages.

This model still does not take into account habitat interspersion and juxtaposition, which probably have significant effects on selection for many species. For example, Porter and Church found that a standard use—availability analysis of habitat selection by wild turkeys indicated an avoidance of agricultural lands, when in reality turkeys used agricultural lands extensively, but only those near hardwood forests. In essence, the turkeys viewed the edge between field and forest as a separate habitat type. Similarly, Neu et al. posited that moose might feed preferentially in a recent burn, but not too far from the surrounding forest. Thus they defined four habitat types—the interior of the burn, the burn periphery, the forest edge adjoining the burn, and the remainder of the forest—and through a simple chi-square analysis showed selection for the edge. Most situations probably are not this simple. Many authors have admitted to the importance, but difficulty, of incorporating spatial aspects of habitats in use—availability analyses. Porter and Church proposed a method whereby the study area is gridded into cells and an assortment of habitat variables within those cells are examined through multivariate analyses to find those that best explain differential use of cells. Litvaitis et al. did just that in a study of bobcats (Felis rufus), which predated the paper by Porter and Church.

Litvaitis et al. looked for associations (using regression and DFA) between the number of radiolocations within 25-ha cells inside home ranges and measurements of several habitat variables sampled there; however, They found that these habitat variables poorly explained variation in frequency of use. Servheen and Lyon used a similar approach in assessing habitat selection by caribou (Rangifer tarandus). They measured habitat variables in 40-ha circles around telemetry locations and sought to find those that best differentiated the areas that the animals used seasonally. Although they had no real measure of juxtaposition or interspersion of habitats, their 40-ha circles contained habitats neighbouring the one actually occupied, so the composition of these circles gave an indication of habitat combinations that corresponded with seasonal use. In another similar approach, Clark *et al.* used grid cells that could encompass several habitat types near the locations of radiocollared black bears.

A suite of habitat characteristics (including the number of different habitat types) within each cell used by bears were combined to form what they called an ideal habitat profile. The habitat quality of each cell in the study area was then assessed by comparing it to this hypothetical ideal cell. Each of these studies looked at differential use, rather than use in terms of availability, and thus avoided the fatal flaw of habitat selection studies. Site attribute studies are like the habitat use studies just discussed, except that instead of comparing cells with varying degrees of use, they categorize cells (sites) simply as used or unused; based on this, important habitat variables are identified. Interspersion and juxtaposition of habitats can thus be investigated. For example, Coker and Capen examined cowbird (Molothrus ater) selection for habitat patches of various size, shape, and location relative to other habitats by entering these variables in a logistic regression with use (used or not used) as the dependent variable. Similarly, Chapin et al. compared habitat variables (including an index of the extent of habitat edge) in grid cells of different sizes that were used (*i.e.*, had at least one telemetry location) by American martens with those in cells not used by martens, and also compared characteristics of forest patches that were used and not used.

McLellan argued that observed use is a better indicator of habitat selection than use relative to availability. He reasoned that an animal familiar with its home range knows the availability and location of resources, so an animal's location at any given moment represents selection. He gave an example of a person at a buffet selecting a slice of beef from a 500-kg steer and an equalsized slice of pork from a 100-kg pig; based on use alone, pork and beef were selected equally, but compared to availability, pork appears to be selected over beef, which is obviously absurd. However, had the steer and pig been cut up in equal-sized chunks and distributed over a large area, and after considerable searching the person still returned with an equal quantity of the two foods, active selection for pork would indeed seem apparent. The key difference is that in the latter case the person had to search for the food; selection was evidenced by the extra effort expended in finding the pork (and apparently bypassing chunks of beef).

This searching for resources is really the basis for the development of use—availability comparisons and explains why it originated with studies of diet. In most cases animals do not know the location of all foods in their home range, so dietary selection based on availability may be appropriate. However, habitats are not spread around like chunks of pork and beef, but occur in large patches, the locations of which are known by the animals; thus habitats are probably more like McLellan's whole steer and whole pig than the cut up chunks of meat spread randomly around. Consider some actual examples of how observed use and use versus availability can lead to disparate interpretations of selection. Prayurasiddhi investigated use and selection among two large ungulates, gaur (Bos gaurus) and banteng (B. javanicus), in Thailand. He differentiated two general study area boundaries,

one of which more closely matched the area that his radiocollared animals used most intensively.

He also used actual home range boundaries as a third representation of the study area and hence the available habitat. He found that this variation in the area considered to be available habitat resulted in drastic differences in perceived habitat selection. One habitat that received 46 per cent of use by gaur was deemed to be selected for, whereas another habitat that received 45 per cent of use was seemingly "selected against." In one case banteng were judged to be unselective in their use of a habitat in which they spent 75 per cent of their time. Prayurasiddhi recognized these difficulties and decided to evaluate seasonal changes in habitat selection and habitat-related differences among species based on use alone, rather than use compared to availability.

In another analogous situation, Macdonald and Courtenay found that crab-eating zorros (foxes, Cerdocyon thous) in Amazonian Brazil spent most of their time (64 per cent) in wooded savanna and scrub habitats; however, because these two habitats were abundant within the home ranges of the study animals, they ranked the lowest in apparent preference (sixth and seventh among seven defined types) based on a comparison of use to availability.

During the wet season, however, when lowland habitats flooded, the zorros used upland wooded savannas and scrub habitats even more, making the apparent preference. Ranking for these rise; in reality, the area of available (not flooded) lowland habitats diminished, but this reduction could not be measured and therefore was not taken into account in the use:availability calculations. The authors realised that the seasonal difference in apparent habitat preferences was thus an artifact of unmeasured changes in availability. In another such case, Garrett et al. found that tidal flats were the principal foraging habitat for bald eagles (Haliaeetus leucocephalus) and observed that nearly one-fourth of their perch sites were within this habitat. That is, based on use alone, this area was clearly attractive to these birds. However, because this habitat was so widely available, especially at low tide, a comparison of use to availability suggested that the eagles avoided it. Apparent preference thus changed radically with tidal fluctuations.

A problem with assessing selection from use alone is that use often does change with availability. In the eagle study, tidal flats were rarely used during high tide because they were largely unavailable. The eagles may have preferred tidal flats to other habitats, but use of these areas depended on availability, up to a point. When availability exceeded some threshold they did not continue to increase their use, and when it declined below some threshold their use of the area ceased. Between these thresholds some relationship may exist between use and availability. However, relationships between use and availability probably vary among habitat types, and within each habitat may depend on what other habitats are available.

It might seem that site attribute studies avert these problems because selection is inferred not from differences between use and availability, but from differences between used sites and other sites. However, the comparative sites are taken proportionately from those available, so the comparison is still, in a sense, the same as in the use—availability design.

Consider a study that locates 10 nest sites in each of two available habitat types and then compares the characteristics of these to 20 randomly chosen points. If one habitat type is much more common than the other, then the 20 randomly chosen points fall mostly within the more common habitat; this sampling could result in possible differences between the average characteristics of the randomly chosen sites and the real nests, suggesting that the species selected for nest sites in the less common habitat type. This logic falters, though, for the reason highlighted in McLellan's example of the buffet, in which a person chose equal portions of beef and pork. In that example, the equal choice of the two types of food was clear evidence of nonselection, or equal preference, regardless of availability.

Accordingly, in the hypothetical case of the 20 nests, one should question why, if there was strong selection for one habitat type, did an equal number nest in each type? The conclusion that this represents evidence of preference for the less common type is (just as in use—availability comparisons) based on the assumption that if individuals had no preference, they would chose sites. Randomly, which is typically thought to mean in proportion to availability. However, suppose each individual, having no preference, just flipped a coin, so to speak, to decide which of the two types of habitats to nest in each year. The result would be as posed in the example, an equal number nesting in each type. There are two other problems (really two aspects of the same problem) with detecting selection using a site attribute design.

James and McCulloch showed that a species could be highly selective for a certain habitat component, as indicated by a low variance among values of this component at the sites it chose, but if the mean value for the component at selected sites approximately equaled the mean in the available habitat (where the variance was much higher than at selected sites), then selection might not be detected (type II error). Conversely, Rexstad et al. tested multivariate statistical procedures that are commonly used in site attribute studies and found that significant results, which in a real study would be indicative of selection for certain habitat components, were often derived from a collection of meaningless data (type I error). Taylor attributed Rexstad et al.'s spurious.

Results to ill-defined hypotheses and inappropriate use of statistical procedures; however, Rexstad et al. showed that their study mimicked the majority of published applications of these techniques. North and Reynolds expanded on Rexstad et al.'s concerns that the statistical procedures used in site attribute studies may yield misleading results because assumptions are commonly violated.

INFERRING HABITAT QUALITY

If habitat selection studies were simply an attempt to better understand behaviour and natural history, errors would merely be a setback in scientific inquiry. However, these studies are typically designed to provide management guidelines. Habitats are managed based on supposed importance in terms of fitness for members of the target population. High-quality habitats, by definition, produce animals with high reproduction and survival, and hence a population with a high growth rate (and high yield, for harvested species). However, the assumption that we can infer habitat quality or suitability from studies of habitat selection—that selection, even if accurately measured, is directly related to each habitat's potential contribution to individual fitness and hence the population's growth rate—represents what the second fatal flaw in the process of habitat evaluation.

One of the best examples of a purported relationship between habitat selection and fitness was provided by a study of leaf-galling aphids (Pemphigus betae) that parasitize narrowleaf cottonwood trees (Populus angustifolia). These aphids overwinter in tree bark, then over a short period in the spring migrate up the tree and become entombed in expanding leaf tissue, where they reproduce. Whitham found that aphids apparently selected large leaves over small leaves, even though large leaves were less available (less than 2 per cent of leaves were over 15 cm; more than 30 per cent were 5 cm or less); moreover, those that colonized large leaves had better survival and reproduction than those colonizing small leaves.

From this evidence it appeared that aphids had the ability to select leaves that offered the highest fitness; if not, the size distribution of colonized leaves would have matched the size distribution of leaves available on the tree (*i.e.*, use would have equaled availability). However, Rhomberg observed that large leaves tended to be those near the tips of twigs (where they receive more light), and these opened a few days later than basal leaves. When the aphids begin their migration, basal leaves are already too mature to enable them to form a gall, so they must colonize the more distal leaves; thus, the basal leaves are in essence nonhabitat. The migration appears to be timed so that the aphids inevitably colonize leaves that are destined to be favourable to their fitness, but individually, they do not select these among less favourable habitats. This is not to say that other animals do not select habitats favourable to their fitness; certainly they do. But if the comprehensive and seemingly compelling data on a simple system such as aphids on cottonwood trees can be misinterpreted, then clearly there is much room for misinterpreting relationships between habitat selection and fitness in more complicated systems. The presumed link between selection and fitness is rarely tested, partly because such tests are difficult, but also because this relationship is viewed more often as fact than as assumption. However, there are many reasons why this relationship may not hold. Habitats that are used infrequently may be more important than suggested by the time spent there.

Conversely, habitats used preferentially for activities that require a lot of time (*e.g.*, resting) may be less important than indicated by their use; a resting habitat may be substitutable, with little effect on fitness, whereas a specific foraging habitat (or a water hole in an arid environment) may be more essential, even if not used very much. Differentiating activity-specific habitat use may help alleviate this difficulty, although this, too, is seldom done, and moreover, does not necessarily circumvent the problem. For example, Powell studied winter foraging of fishers, which preyed heavily on porcupines (Erethizon dorsatum) but spent a disproportionately small amount of time hunting in upland hardwood habitats where porcupines were common.

The reason, he found, was that fishers rapidly located porcupines at known den sites, thus minimizing their search and chase times. In contrast, fishers had a harder time hunting snowshoe hares, so they spent a larger amount of time in lowland conifer habitats where hares were more common. Clearly, the relative importance of the two habitats to fishers was not reflected by their time spent hunting in each. For animals that feed on a variety of different foods, a mixture of different habitats may be more beneficial in many respects than a single, highly preferred type, and the time spent in each may be a poor indicator of importance of either the specific type or the overall mix. Omnivorous Egyptian mongooses (Herpestes ichneumon), for example, favour one specific habitat for resting, but use a mix of habitats when feeding. Fitness also may be affected by predation and interspecific or intraspecific competition. These factors can change the cost:benefit ratio of a habitat and hence alter an animal's habitat use. Cowlishaw found that baboons (Papio cynocephalus) spent more time feeding in a habitat with poor food but a low risk of predation than in a food-rich habitat with a high risk of predation. Fitness was probably enhanced by this choice of habitats.

Fitness certainly was enhanced for a herd of caribou in Ontario, Canada, that spent most of the time on an island. The nearby mainland had higherquality forage but also a high density of wolves. Ferguson et al. found that the island occupants sacrificed nutrition, which was reflected in smaller body and antler size as well as starvation of their calves, for overall higher survival, which enabled this population to persist while herds that remained on the mainland perished from predation. One could argue that the animals on the mainland made poor choices in habitat selection. Animals may not always correctly perceive risks, especially in a novel or changing situation, so their choice of habitats may not necessarily be best in terms of their fitness. Pollution is one such risk that is rarely evaluated in habitat studies. Consider Mallory et al.'s study of the effects of acid rain on habitat quality for common goldeneyes (Bucephala clangula).

Acid rain killed fish in lakes, which enabled proliferation of invertebrates, thus providing more food and enhanced reproduction for goldeneyes. However, these ducks may inevitably suffer other consequences of the acidified habitat. In some situations the choices made by animals may maximize fitness over the long term (*e.g.*, if the changes are ephemeral); in other cases animals may require time to adapt. Suppose that an animal uses alternate habitats more often when the normally preferred habitat becomes less profitable because of increased density and hence competition for resources. If the alternative habitats are more available than the preferred habitat, then use of these alternative habitats can end up higher than that of the otherwise preferred habitat. Situations like this seemingly justify considering habitat use in terms of availability. However, Hobbs and Hanley showed both intuitively and with computer modeling that intraspecific competition may alter use: availability ratios such that they may not reflect the true quality of a habitat in terms of its potential contribution to population growth.

Certainly interspecific competition and predation are integral components of habitat quality, and managers concerned about a given species should not evaluate habitats independent of these factors. However, perceptions of habitat quality that are affected by density of the targeted species are an artifact of the method. The habitat preference should be innate, and intricately tied to fitness; if our perception of preference declines as the preferred habitat becomes more densely populated, then our perception is clearly mistaken.

Hobbs and Hanley concluded that "use/availability data inherently reflect differences in animal density among habitats" (independent of social behaviour) and therefore reveal little about habitat quality. Consequently, they recommended abandonment of the use—availability design in favour of an approach that enables a direct investigation of the link between habitats and fitness. Site attribute studies often focus on sites of biological importance (as reviewed earlier) and therefore may provide more direct insights into habitat variables that affect fitness. However, as pointed out by North and Reynolds, conclusions from studies of this sort are based on two fundamental, but questionable, underlying assumptions: that used sites are in suitable habitat and that unused sites are in unsuitable habitat.

Consider the study by Meyer *et al*, who observed significant habitat-related differences among sites that were occupied versus unoccupied by spotted owls (Strix occidentalis). They found, as expected, that these owls tended to select sites having a large proportion and large patches of old-growth forest, but suggested that the owls could tolerate certain amounts of clearcut and young forest (based on the fact that used sites contained such areas). Their reproductive data were inconclusive about what the owls could tolerate from a population standpoint, but Meyer et al nonetheless asserted "that resource selection probability functions are a reliable tool for assessing capability of landscapes to support northern spotted owls." Hence they concluded that their habitat selection model "can be used to predict the probability that a given landscape mosaic will be a suitable spotted owl site" (emphasis mine). To avoid assumptions related to suitability in site attribute studies, North and Reynolds proposed that instead of comparing habitat characteristics of used sites with unused or random sites, used sites be categorized by their intensity of use (*e.g.*, 1–2 per cent of locations = low use, 3–10 per cent medium use, more than 10 per cent = high use), and habitat characteristics compared among such categories (*i.e.*, excluding unused sites) with a form of logistic regression.

Thus this is not really a site attribute design (according to my definition), nor is it even applicable to the types of situations in which site attribute studies are commonly used (*i.e.*, in which an animal remains at a single site for a fairly long period of time). Rather, this procedure represents just another means of investigating habitat characteristics associated with differing degrees of use. Whereas these methods may avoid many of the statistical and general assumptions of site attribute studies, they digress further from the issue of habitat-related fitness because the studied sites, instead of being of some special biological significance, are simply points where the animal was located some number of times. As North and Reynolds recognized, intensity of use may not be a good indicator of habitat quality.

DEMOGRAPHIC RESPONSE DESIGN

The best measure of habitat quality would be a test of its effects on demographic parameters such as population growth and carrying capacity. Such tests are extraordinarily difficult in most situations, as evidenced by the scant published studies of this nature. Because habitat-specific density is actually a reflection of differential habitat use, investigations of habitat-related density suffer the same drawbacks as studies of habitat use. Density tends to be an ineffective measure of habitat quality because it may fluctuate widely, is subject to sizable errors in estimation, and may be largely influenced by social factors. Van Horne gave several examples of situations in which juveniles were restricted from settling in the best habitats and thus accumulated in large numbers in poorer quality habitats. She indicated that such circumstances are likely to be common among generalist species with high reproductive rates and a social hierarchy. For these species in particular, then, habitat-specific density would probably be a poor indicator of habitat quality unless the population is well below carrying capacity. A good example was provided by Messier *et al.*, who showed that density of muskrats (Ondatra zibethicus) during a general population increase swelled 30- to 90fold in low-quality habitats but much less in high-quality habitats.

Considering that, in general, animals in poor-quality habitats should be trying to leave and ones in high-quality habitats trying to stay (and keep competitors out), Winker et al. posited that turnover rate would be a better index of habitat quality than density. They measured turnover rates for wood thrushes (Catharus mustelinus) by examining recapture rates and telemetry movement data; low-quality habitat was defined as that in which recapture rates were low and many radiotagged birds were transient visitors. They found density and habitat quality, assessed in terms of turnover rates, to be inversely related. Notably, Winker et al.'s turnover rate model may not be applicable to other species, even other territorial, noncolonial songbirds, some of which are preferentially attracted to habitats occupied by conspecifics, which they use as a cue to habitat quality.

Conspecific attraction tends to perpetuate use of the same areas across generations, so even if habitat quality deteriorates, high densities may be maintained through tradition. Other competing species or unidentified confounding variables also may weaken the linkage between habitat quality and density. Maurer measured density and various habitat characteristics for five species of grassland birds; the habitat models developed to explain species-specific density in one study area were inexplicably poor predictors of density in a nearby area with similar habitat. Kellner et al. found that density was positively related to reproductive success in only 7 of 17 bird studies that they reviewed; more studies showed a negative relationship. Sherry and Holmes felt that density should be relied on as an indicator of habitat quality only if it is corroborated by other data, as was the case in their study, in which population density and weight loss of wintering migrant birds were correlated (high weight loss in areas with low densities) and both were related to habitat type.

Reproduction and survival data may be more apt to reflect real influences of habitat on demographics. However, reproduction and survival are probably also tied to habitat in a complex manner. For example, a number of studies observed a direct relationship between cover and the survival (and thus density) of voles (Microtus spp.), but a lower threshold exists below which reductions in cover have little effect on vole density; above the threshold, survival and density increase but eventually reach an upper asymptote. The vole studies found that cover provides food as well as protection from predators, and also may affect microclimate, activity patterns, and interactions among conspecifics, all of which affect the cover—density relationship. Moreover, male and female voles have different responses to varying cover, and cover—demographic relationships tend to be different for other small grassland rodents.

Each of the various habitat components that relate to an animal's fitness probably has thresholds, asymptotes, and inflection points, and these limits may vary with the mix, shape, size, and juxtaposition of habitat components available; however, few attempts have been made to assess any of these factors individually, let alone in combination. Several studies also found that density-dependent effects may reduce reproduction or survival independent of habitat quality or may even result in higher fitness in low-quality, less crowded habitats. Zimmerman found that nesting success of dickcissels (Spiza americana) did not differ between habitats and was unrelated to density, but females nested preferentially in the habitat preferred by males; males chose this more heterogeneous habitat because they could sequester more nest sites and thus mate with more females. In other studies, reproduction and survival were found to be unrelated to measured habitat variables, despite evidence of habitat selection, possibly because of confounding effects of weather, human disturbance, measurement error, and other factors. Even experimental demographic response studies have been plagued with unexpected variations in confounding variables.

Nevertheless, some carefully designed studies observed links between habitat and reproduction or survival and through further investigation discovered the underlying causes. A final major problem, discussed previously in relation to use—availability and site attribute designs, is scale. Levin showed clearly that there is no single correct scale for studying ecological relationships. Animals view and react to their environment at various scales. Human perceptions of ecological systems are inescapably biased or incomplete because they are filtered by the observational scale chosen for the investigation.

Some demographic response studies have recognized this and have adopted a multiscaled approach. For example, Orians and Wittenberger found that densities of yellowheaded blackbirds (Xanthocephalus) were higher on marshes with higher food (insect) abundance but that density of blackbird territories within these marshes was related to vegetational structure, not food. The authors postulated that these birds may not select food-rich territories because they often hunt outside their territories. Also, they establish their territories before the full emergence of insects and hence may not be able to predict future food abundance on a scale smaller than the marsh level.

Pedlar et al. developed habitat models on two different scales to explain variation in raccoon (Procyon lotor) density. One model was fit to macrohabitat features and the other to microhabitat variables, after which the two were combined to form a more comprehensive model relating density to habitat at both scales. Morris observed both macrohabitat and microhabitat differences among several species of small mammals, suggesting habitat selection on both scales, but found that variation in density was much more evident at the macrohabitat level. On a larger scale, Dooley and Bowers discovered, counter to their expectations, that densities and population growth rates of voles were higher in patches within a fragmented landscape than in an unfragmented landscape, whereas total population size was higher in the unfragmented landscape (because more total habitat was available). Landscape fragmentation caused overall habitat loss but, on a finer scale, enhanced reproduction within individual habitat fragments. Similarly, Brown and Litvaitis noted that some mammalian predators that hunt preferentially in forests nonetheless often exist at lower densities in homogeneous forests than in forests interspersed with disturbed areas. Each of these examples represents cases in which a single-scale investigation would have failed to detect habitat variables affecting population demography. The previous examples all concerned spatial scale. Time scale may be equally important. The demographic value of a habitat may become evident only in the long term, after a population has been subjected to the stresses of a periodic drought, severe winter, or failed food crop.

APPLICATIONS AND RECOMMENDATIONS

A great deal of effort continues to be invested in habitat-related studies of wildlife. In the United States, federal land management agencies in particular have focused on developing formalized procedures for evaluating habitat for wildlife. The procedures that have been adopted rely on models derived by species experts, who in constructing these models tend to rely more on experience than on empirical data. Therefore, the models are really hypotheses in need of testing.

However, because these models hypothesize explicit relationships between habitat attributes and animal populations (so-called habitat suitability indices), they cannot be rejected or accepted in normal scientific fashion; that is, none of the relationships are likely to be exactly correct. Thus it seems inappropriate to suggest, as is common parlance, that they should (or even could) be "verified" or "validated." Several attempts have been made to test such models. Often this process is circular, involving just another panel of experts making qualitative assessments. In other cases, models have been tested using results of a study on habitat use or use relative to availability, with the inevitable associated shortcomings. In some instances, habitat management prescriptions based on "common knowledge" or expert opinion have, through collection of better data, been proven faulty. One case in which model-derived habitat scores for individual home ranges were compared with reproduction, juvenile growth rates, and home range size, but no significant relationships were observed. Often, models have been tested by comparing habitat-specific densities to model predictions.

However, even if a model explains a significant portion of the variation in density, the data collected to test (or purportedly validate) the model could be better used to modify it or build new one. In most cases, habitat models have proved to be poor predictors of animal density, indicating either defectiveness of the model, lack of a clear habitat—density relationship, or effects of other confounding factors, such as hunting pressure, which are also habitat-related. Bender et al. found that if the variance around the estimated values of the model inputs were taken into account (a process that is not commonly done), suitability scores for a variety of habitats that appeared very disparate were not significantly different; that is, the parameter estimates were not precise enough to even enable the model to be tested.

A major inherent but generally unstated (maybe unrecognized) assumption of habitat suitability models is that high-quality habitats (*i.e.*, habitats that confer high fitness) are in fact suitable. Explicit tests of this assumption are rarely conducted, yet counterexamples exist. In one case, a habitat suitability model for Florida scrub-jays (Aphelocoma coerulescens) correlated well with demographic performance, but most of the area was found to be a population sink where mortality exceeded reproduction. Apparently, as birds competed to occupy the best habitats, they were less alert to predators, and thus suffered high mortality. In another example, Kirsch found that interior least terns (Sterna antillarum, least terns nesting in noncoastal areas) selected high-quality nesting habitats, but possibly because of disturbance, their productivity was not sufficient to maintain population size (*i.e.*, the nesting habitats were unsuitable). Lomolino and Channell observed that remnant populations of endangered mammals often occur near the periphery of their former ranges; because the periphery of the range represents the edge of suitable habitat, studies of habitat suitability of endangered mammals based on habitat use in existing populations are likely to be misleading. These are not gratifying results.

It has not been for lack of effort, expense, or analytical developments that relationships between habitats and population growth often elude detection. It is simply the complexity of the interactions between animals and their environment that make such relationships deceptively difficult to understand. A case in point is the spotted owl, which has undergone intense scrutiny because of its threatened status and apparent proclivity for old-growth forest in a region where the economy is tied largely to timber; despite a plethora of habitat studies, significant debate persists among ecologists as to the critical habitat requirements of this species, and why it prefers old-growth forest. However, substantive flaws in the most commonly used techniques for studying wildlife—habitat relationships apparently have not been widely recognized. It is time to reconsider the ways these techniques are used in evaluating habitat quality. Studies of the use of habitat have merit, but also many limitations.

Bibliography

Ager, D.: *The Nature of the Stratigraphic Record*, London: Macmillan Press, 2003.

Alexander, R.: *The Chordates*, London: Cambridge University Press, 2001.

Andrewartha, H.: *Introduction to the Study of Animal Populations*, London: Methuen Press, 2000.

Arnold, C.: *Introduction to Paleobotany*, New York: McGraw-Hill Press, 2006.

Barrington, E.: *The Biology of the Hemichordata and Protochordata*, Edinburgh: Oliver and Boyd Press, 2003.

Bertalanffy, L.: *General Systems Theory: Foundations, Development, Applications*, New York: Braziller Press, 2004..

Black, R.: *The Elements of Paleontology*, London: Cambridge University Press, 2001.

Brouwer, A.: *General Paleontology*, Chicago: University of Chicago Press, 2005.

Calvin, M.: *Chemical Evolution*, London: Oxford University Press, 2000.

Carlquist, S.: *Island Life: A Natural History of the Islands of the World*, New York: Natural History Press, 2004.

Chorley, R.: *Physical Geography: A Systems Approach*, London: Prentice-Hall, Press.

Dasmann, R.: *Environmental Conservation*, New York: Wiley Press, 2001.

Dobzhansky, T.: *Evolution, Genetics and Man,* New York: John Wiley & Sons Press, 2003.

Ebbesson, S.: *Comparative Neurology of the Telencephalon,* New York: Plenum Press, 2005.

Feller, W.: *An Introduction to Probability Theory and its Application,* New York: Cambridge University Press, 2000.

Flint, R. F.: *Glacial and Pleistocene Geology,* New York: John Wiley & Sons Press, 2005.

Garner, H.: *The Origin of Landscapes: Synthesis of Geomorphology,* New York: Oxford University Press, 2001.

Glass, B. P.: *Introduction to Planetary Geology,* Cambridge: Cambridge University Press, 2004.

Hallam, A.: *Jurassic Environments,* Cambridge: Cambridge University Press, 2003.

Le Gros Clark: *The Fossil Evidence for Human Evolution,* Chicago: University of Chicago Press, 2001.

Moodie, R.: *Paleopathology: An Introduction to the Study of Ancient Evidences of Disease,* Urbana: University of Illinois-Urbana Press, 2004.

Nelkin, D.: *Science Textbook Controversies and the Politics of Equal Time,* Cambridge: M.I.T. Press, 2002.

Romer, A.: *Vertebrate Paleontology,* Chicago: University of Chicago Press, 2001.

Sherrington, C.: *The Integrative Action of the Nervous System,* Cambridge: Cambridge University Press, 2002.

Simpson, G.: *The Major Features of Evolution,* New York: Columbia University Press, 2003.

Sirks, M.: *The Evolution of Biology,* New York: Ronald Press, 2003.

Index

G

I

M

O

P

R

S

W

Z